William Webb
The 5G Myth

William Webb
The 5G Myth

When Vision Decoupled from Reality

Third Edition

ISBN 978-1-5474-1728-5
e-ISBN (PDF) 978-1-5474-0118-5
e-ISBN (EPUB) 978-1-5474-0120-8

Library of Congress Control Number: 2018956918

Bibliographic information published by the Deutsche Nationalbibliothek
The Deutsche Nationalbibliothek lists this publication in the Deutsche Nationalbibliografie;
detailed bibliographic data are available on the Internet at http://dnb.dnb.de.

© 2019 William Webb
Published by Walter de Gruyter Inc., Boston/Berlin
Printing and binding: CPI books GmbH, Leck
Typesetting: MacPS, LLC, Carmel
Cover Image: Vertigo3d / iStock / Getty Images Plus

www.degruyter.com

myth
n.

A belief or set of beliefs, often unproven or false, that have accrued around a person, phenomenon, or institution.
(Farlex, *The Free Dictionary*, https://www.thefreedictionary.com/myth)

About De|G PRESS

Five Stars as a Rule

De|G PRESS, the startup born out of one of the world's most venerable publishers, De Gruyter, promises to bring you an unbiased, valuable, and meticulously edited work on important topics in the fields of business, information technology, computing, engineering, and mathematics. By selecting the finest authors to present, without bias, information necessary for their chosen topic *for professionals*, in the depth you would hope for, we wish to satisfy your needs and earn our five-star ranking.

In keeping with these principles, the books you read from De|G PRESS will be practical, efficient and, if we have done our job right, yield many returns on their price.

We invite businesses to order our books in bulk in print or electronic form as a best solution to meeting the learning needs of your organization, or parts of your organization, in a most cost-effective manner.

There is no better way to learn about a subject in depth than from a book that is efficient, clear, well organized, and information rich. A great book can provide life-changing knowledge. We hope that with De|G PRESS books you will find that to be the case.

Acknowledgments

Because the ideas in this book are unconventional, I have sought critical reviews from a range of colleagues. A series of debates arranged through Cambridge Wireless were most helpful, and I would like to thank the panelists for their insights—namely, Joe Butler, Moray Rumney, Paul Ceely, Howard Benn, Richard Feasey, Tony Lavender, and Stephen Howard, as well as the chief executive officer of Cambridge Wireless, Robert Driver, for arranging the debates. Others who have kindly reviewed material include Steve Methley, Brian Williamson, Dennis Roberson, and Rob Kenny. Of course, this in no way implies that they agree with everything written here; indeed, sometimes it is quite the converse.

I have been informed and inspired by a wide range of colleagues who have presented at conferences or worked with me on the many projects that helped build the insight presented here. Thanks to all.

And finally, to my family for putting up with my introspection as I pondered the wisdom of writing this book.

About the Author

William Webb is the director at Webb Search Consulting, a company specializing in matters associated with wireless technology and regulation. He is also the chief operating officer of Weightless SIG, the standards body that is developing a new global machine-to-machine (M2M) technology. During 2014–2015 he was president of the Institute of Engineering and Technology (IET), which is Europe's largest professional engineering body.

In 2011, Mr. Webb was one of the founding directors of Neul, a company developing M2M technologies and networks which was subsequently sold to Huawei in 2014. He has also been a director at the UK's Office of Communications (Ofcom), where he managed a team providing technical advice and performing research across all areas of Ofcom's regulatory remit. During his work with Ofcom, he led some of the major reviews, including the *Spectrum Framework Review*, the development of spectrum usage rights, and cognitive or white space policy. William has also worked for a range of communications consultancies in the United Kingdom in the fields of hardware design, computer simulation, propagation modeling, spectrum management, and strategy development and spent three years providing strategic management across Motorola's entire communications portfolio in Chicago.

William has published 16 books and over 100 papers and holds 18 patents. He is a visiting professor at Surrey and Southampton Universities; an adjunct professor at Trinity College Dublin; and a fellow of the Royal Academy of Engineering, the Institute of Electrical and Electronics Engineers (IEEE), and the IET. In 2015 he was awarded the honorary degree of doctor of science by Southampton University in recognition of his work on wireless technologies and honorary doctor of technology by Anglia Ruskin University in recognition of his contribution to the engineering profession. His biography is included in multiple *Who's Who* publications around the world. William has a first-class honors degree in electronics, a PhD, and an MBA. He can be contacted at wwebb@theiet.org.

1. W. Webb, *Our Digital Future* (Amazon, 2017).
2. M. Cave and W. Webb, *Spectrum Management: Using the Airwaves for Maximum Social and Economic Benefit* (Cambridge, UK: Cambridge University Press, 2015).
3. W Webb, *Dynamic White Space Spectrum Access* (Cambridge, UK: Webb Search Limited, September 2013).
4. W. Webb, *Understanding Weightless* (Cambridge, UK: Cambridge University Press, March 2012).
5. W. Webb, *Being Mobile: Future Wireless Technologies and Applications* (Cambridge, UK: Cambridge University Press, 2010).
6. M. Cave, C. Doyle, and W. T. Webb, *Essentials of Modern Spectrum Management* (Cambridge, UK: Cambridge University Press, 2007).
7. W. T. Webb, *Wireless Communications: The Future* (Chichester, UK: Wiley, 2007).
8. L. Hanzo, K. Ng, T. Keller, and W. Webb, *Quadrature Amplitude Modulation*, 2nd ed. (Chichester, UK: Wiley, 2004).
9. W. T. Webb, *The Future of Wireless Communications* (Norwood, MA: Artech House, 2001).

Further books and a full list of publications are available at www.webbsearch.co.uk.

Contents

Chapter 1: Introduction to 5G —— 1
    The Role of the Mobile Network Operator —— 1
    Next Generation Cellular —— 2
    The 5G Themes —— 3
    Opinions Vary on 5G —— 4
    5G Technology —— 4
    5G Radio Spectrum —— 5
    What the Consumer Might Expect —— 5

Chapter 2: Learning from Prior Generations —— 7
    A Quick Trawl Through History —— 7
    Moving from 1G to 2G —— 9
    Moving from 2G to 3G —— 10
    Moving from 3G to 4G —— 11
    New Entrants into the Game —— 12
    Extrapolations Do Come to an End —— 13

Chapter 3: Can Demand Grow Indefinitely? —— 15
    Volume Versus Speed —— 15
    What Speed Is Needed —— 15
    Latency —— 18
    Predictions of Volume —— 20
    Internet of Things —— 27
        What Is the World of Connected Machines? —— 27
        The Current Position —— 29
        Which Applications Will Lead? —— 30
        Privacy and Security —— 31
        Where Might It End Up? —— 31
        Implications for Network Load —— 32
    Could Anything Make a Difference? —— 32
    Conclusions —— 34

Chapter 4: Technology Reaches Its Limits —— 37
    MIMO —— 38
    Small Cells (Microcells) —— 41
    mmWave —— 44
    Full Duplex —— 47
    Core Network Evolution —— 47

Heterogeneous Networks —— 48
A Route to Structural Change —— 50
Conclusions —— 51

Chapter 5: Economics Reaches Limits —— 53
MNO Performance —— 53
5G Costs —— 58
Expectations of ARPU Growth —— 59
New Verticals —— 62
Conclusions —— 62

Chapter 6: Why Key Players Are Enthusiastic —— 65
Academics —— 65
Equipment Supply Industry —— 66
Mobile Network Operators —— 67
Government —— 68
Outlandish Claims —— 69
The Major Players Cannot Be Wrong —— 70
Conclusions —— 70

Chapter 7: The 5G Vision —— 73
A Collection of Visions —— 73
NGMN View of Potential Applications —— 75
  Broadband Access —— 75
  Higher User Mobility —— 77
  Massive Internet of Things —— 78
  Extreme Real-Time Communications —— 80
  Lifeline Communication —— 81
  Ultrareliable Communications —— 82
  Broadcast-Like Services —— 84
  Summary of NGMN Requirements —— 85
Conclusions —— 85
Chapter 7 Appendix: Autonomous Vehicles in More Detail —— 87

Chapter 8: Alternative Futures —— 91
Why Consistency Is More Important than Speed —— 91
How to Deliver Consistency —— 92
  Trains —— 92
  Rural Areas —— 94
  In the Home —— 95

    In the Office —— 96
    Public Buildings —— 96
    Dense Areas —— 96
    Summary —— 97
A "Wi-Fi First" World —— 97
    Making and Receiving Calls over Wi-Fi —— 99
    Automated Passwords —— 99
    Security —— 101
    Reliance on Unlicensed Spectrum —— 101
    Failure of Municipal Wi-Fi —— 102
    5G and Wi-Fi —— 102
    Summary —— 103
Regulatory and Governmental Action —— 103
    Policies No Longer Needed —— 104
    New Policies Needed —— 104
Delivering the Internet of Things —— 105
An Alternative Future —— 106
    2018 —— 106
    2019 —— 107
    2020 —— 109
    2021 —— 109
    2022 —— 110
Conclusions —— 111

Chapter 9: Regulation, Competition, and Broadband —— 113
Regulation and Competition —— 113
Radio Spectrum —— 116
Regional Differences —— 117
Broadband Fixed Access —— 119
Conclusions —— 121

Chapter 10: How the Future Plays Out —— 123
Why 5G As Currently Envisioned Is Flawed —— 123
A Better Vision: Consistent Connectivity —— 124
Significant Industry Structural Change —— 125
5G Becomes Whatever New Stuff Happens —— 126
The Future Is Bright—Once the Vision Is Realigned —— 128

List of Abbreviations —— 129

Index —— 131

Preface

The first edition of this book was published in November 2016—a time when the anticipation and expectation from 5G was enormous. The term "5G" permeated almost every major announcement from the wireless telecommunications industry; governments, politicians, and international bodies all wanted to be associated with the future vision it promised. The book attempted to bring some pragmatic reality to a world where hype and optimistic thinking appeared to predominate.

Nearly two years have passed since then—a long time in the development of a new cellular technology, especially one that in 2016 was forecast to be deployed by 2018. More evidence has come to light regarding 5G, and I have engaged in much debate about my thinking that has helped further develop my arguments. As a result, it seems timely to provide an update to the book, albeit a relatively minor one. This third edition provides a new introduction to 5G in Chapter 1, and the book has been made more readable for a broader audience without sacrificing any of the content.

Between the first and this edition, very little has actually happened, much as I predicted. Although 5G specifications have progressed and some elements are now finalized, much remains to be done. Trials and test-beds apparently continue (although there is very little information as to what they have discovered), and the term "5G" still pervades many announcements. No new uses or services have emerged which might change the average revenue per users (ARPUs) or user requirements, and no insights are yet available from efforts to use mmWave solutions to deliver fixed-wireless access. However, one important change *has* occurred—skepticism in 5G has grown. In November 2017 Vodafone's chief technology officer (CTO) stated that 5G was overhyped and the key benefit was efficiency gains,[1] while BT's chief executive officer (CEO) said that neither he nor many of the other CEOs he talked to could find a business case for 5G.[2]

As I predicted, ARPUs have remained flat, or in some cases tended downward—in the United States, ARPUs have fallen by a massive 7 percent in the last year. There are early signs that growth in data usage may be slowing. The British Broadcasting Corporation (BBC) recently reported on a study that teenagers were, for the first time, using their phones slightly less. Reports from consultants show

---

[1] James Davies, "We Need to Stop Talking 5G BS—Vodafone CTO," *telecoms*, November 15, 2017, http://telecoms.com/486130/we-need-to-stop-talking-5g-bs-vodafone-cto/.
[2] James Davies, "We're Struggling with the 5G Use Case Right Now—BT CEO," *telecoms*, November 16, 2017, http://telecoms.com/486156/were-struggling-with-the-5g-use-case-right-now-bt-ceo/.

that in some countries—notably Japan, Sweden, and Singapore—data growth has declined to minimal levels. There is still a long way to go before we can be sure that we are starting to see the end of the growth phase, but the signs are pointing in that direction.

It seems clear that 5G is not going to be deployed in the short term—at least not a 5G that conforms to 3rd Generation Partnership Project (3GPP) standards. Recent responses to a consultation by the regulator in Singapore seemed to indicate 2020 is now more widely viewed as the most likely date of introduction, with some tending toward 2025. The Global System for Mobile Communication Association (GSMA) recently predicted that only 40 percent of the global population would have 5G coverage by 2025, and Ericsson has stated that they did not expect to see 5G revenues before 2020 (which means networks ready for consumer use after 2020). Recent presentations from mobile operators suggest a very gradual and incremental deployment; some congested 4G cells have had minor upgrades to enable "5G capabilities" that are really just evolved 4G that may expand to full 5G base stations sometime in the future. The rush to 5G and desire for 5G leadership appears to have abated somewhat.

Equally, not that much has happened with alternative networks like Wi-Fi. Commercial networks blending Wi-Fi and cellular have not gained much traction, although there have been myriad government schemes to deploy public systems. For example, the Indian regulator Telecom Regulatory Authority of India (TRAI) has recently launched an open public Wi-Fi framework to allow hotspot owners to offer service to mobile phone users; the European Union is funding and promoting a continent-wide deployment; and Singapore continues to develop its nationwide Wi-Fi service.

With operators planning to deploy 5G only in congested urban areas, on existing cell sites, and in a mode reliant on 4G for signaling, little will change in the short term. The 4G networks will continue to be the work-horse of cellular and indeed will continue to be enhanced.

Structure of the Book

This book critically examines the myriad proposals for the next generation of mobile communications—5G—and shows that there are many flaws in what is proposed. Indeed, it concludes that the vision of 5G as currently laid out is so badly flawed that it is highly unlikely to be widely implemented. The book does so in a number of stages.

Chapter 1 is an introduction to 5G and related technologies. The idea is to get all readers on the same page so that they will understand the discussion in this book.

Chapter 2 examines the lessons of history, looking back at the transitions through previous generations and showing what simple extrapolations of trends would predict for 5G. We discuss that if previous trends were followed, 5G would become widely deployed in 2022, delivering realistic end-user data rates of 200 megabits per second (Mbps) and an increase in capacity of about two times the level of current networks.

Chapter 3 examines whether such increases in speed and capacity are needed. It shows that users do not value speeds above those already widely available on 4G. It demonstrates that while data requirements are currently growing rapidly, the rate of growth is slowing and, if extrapolated, will result in a plateau in data rate requirements around 2027, with little growth in the 5G era. Hence, it concludes that the advances in speed and capacity that 5G might bring are not needed.

Chapter 4 considers whether the technology is available to provide such gains. It demonstrates that further capacity improvements are very difficult and likely to be expensive to realize, raising the cost of provision for mobile network operators (MNOs). Speed improvements are only available in very high-frequency bands that are limited to dense urban areas and also have high costs. Hence, there are no easy gains from technological improvements. It also suggests that some changes to the core network may have unexpected side effects of enabling different industry structures, splitting the functions provided by MNOs across multiple players.

Chapter 5 considers industry economics and shows that MNOs are in a position of declining revenues relative to gross domestic product (GDP) with a profitability of only half the average across all industry sectors. Few expect this to change with 5G, with the result that investment is highly unattractive unless revenue growth can be stimulated through the delivery of new services.

Chapter 6 asks: why, if the position is so bleak, does the industry collectively appear so supportive of 5G? I discuss that it is not in the interest of any of the key players to cast doubt on a bullish 5G vision and that, for some, the emergence of 5G is essential to their very survival.

Chapter 7 examines the visions set out for 5G in more detail. It shows that they are often unachievable. The collective vision of the MNOs, as set out by the Next Generation Mobile Networks (NGMN) organization, is examined in detail, and each service is shown either to be deliverable via existing wireless solutions such as 4G or to be economically unviable. This chapter explains that the current visions are flawed and that their breadth and lack of reality adds confusion. Few people know what 5G actually is.

Chapter 8 asks what might transpire in place of the current 5G vision. It suggests that consistent connectivity of nearly 10 Mbps everywhere is a more compelling vision and shows how it can be delivered via a mix of 4G and Wi-Fi. The chapter also lays out a possible path for this type of connectivity's introduction, showing how this could result in seismic changes to the structure of the industry.

Chapter 9 looks more broadly at the communications world, considering the impact of regulations, spectrum, and broadband access to the home. It shows that regulation is set to maintain the status quo, which early chapters have shown to be unsustainable, and predicts this will cause further problems for any 5G vision.

Chapter 10 summarizes the findings of the book, sets out why the vision of 5G as currently promulgated by major players is a myth, and discusses what is more likely to transpire. The chapter demonstrates how a world of connected devices, enhanced coverage, a myriad of new applications, and greater productivity can be achieved more quickly and cost-effectively than with the current vision.

Chapter 1
Introduction to 5G

The Role of the Mobile Network Operator

Almost all of us have a mobile phone and are used to being able to talk or access the internet wherever we are. To enable this, mobile network operators (MNOs) such as Verizon and Telefonica build mobile phone networks across the country. These consist of thousands of masts or towers (in the United States, an operator such as Verizon might have about 80,000), each of which transmits a radio signal that can be received by the phones. These masts then connect back into the operator's network, routing the signals to their destination and performing functions such as billing and location tracking.

Most countries have three or four mobile operators who compete with each other to attract subscribers. This competition drives operators to improve their coverage, to provide new features and services, and to keep their prices as low as possible. The rewards for performing well are substantial: in the United States, monthly revenue for basic mobile service is about $40; the larger operators have 150 million subscribers with annual revenue of about $6 billion. But the costs of delivering service are also substantial and include rental fees on towers, costs for upgrading equipment, staff costs, costs for acquiring radio spectrum, repairs, and more.

MNOs are now facing the decision as to whether to upgrade their networks to the latest technology—5G. On one hand, an upgrade is costly; operators would prefer not to spend more on equipment. On the other, if they can deliver a more attractive service to consumers due to the new technological generation, then operators may gain subscribers and hence, extra revenue. This book explores the issues behind that decision.

Mobile, or cellular, is unique in the world of consumer technology. It has evolved in generations. About every 10 years a new generation of mobile technology is announced, whereas in other industries—such as the internet or laptop computers—evolutionary trends appear gradually and incrementally. Mobile is different to some degree because the network is purchased by one entity—the mobile operator—whereas the handset (mobile phone) is purchased by another—the consumer. For them to work together, they need use the same technology; the synchronization provided by the decade-long innovation cycle enables this. But this shared relationship is far from essential. Consumers tend to replace handsets around every two years, so new standards can easily be adopted as needed.

Regardless, the reason for using "generations" appears to have more to do with the industry's reliance on it, so it continues to be preserved—academics thrive on the need for future research; equipment manufacturers rely increasingly on the sales boost a new generation provides; and governments promote their national leadership in technology innovations.

It is not even clear what constitutes a new generation. Mobile communications rely on standards that ensure that network equipment from one manufacturer will work with handsets from another. Standards are developed by an international body called the 3rd Generation Partnership Project (3GPP), who releases updates about once a year. The 3GPP decides whether an update is an "evolution" of one generation or a new generation. Often the step from one generation to another is unclear and further blurred by MNO marketing teams who make claims for next generation deployment to gain a competitive advantage.

It would certainly be possible for mobile communications to thrive without generations. Instead, mobile technologies would gradually evolve, as previous generations have, and consumers would gain from this evolution each time they upgraded their handsets. The difficulty and risk of launching a new generation has sometimes resulted in key players claiming that this is the last time there will be a major generational change. But the next generation—5G—is now almost upon us. Will this be the last generation? Will cellular move toward gradual evolution like other industries?

Next Generation Cellular

Cellular technology has evolved in generations. New generations appear every decade and typically add new features or capabilities: 2G was all about moving to digital communications to improve quality and security; 3G was created to support the move to data, enabling data rates of over 1 Mbps; and 4G was about faster data, with radio solutions capable of around 50 Mbps and networks optimized for data solutions.

A new generation typically requires new radio equipment to be installed by operators, often in a new radio spectrum. For subscribers it means that new handsets are needed to enjoy the benefits of a newer generation. With 4G nearly eight years old, attention has now turned to 5G.

However, the move from one generation to another is not so clear-cut. Generations are progressively upgraded during their lifetime, with these upgrades often termed "3.5G," or something similar. The new generation can sometimes appear very similar to the latest upgrade of the older generation. New generations also tend to be a collection of features, not all of which are introduced initially, so whether a new deployment is fully the next generation or only partially so can be unclear. Marketing pressure to claim leadership tends to compound this issue.

The 5G Themes

Each generation has had a theme, such as faster data. For example, the theme of 4G was faster mobile broadband. There are multiple themes for 5G; these are often described as:
- *Enhanced mobile broadband (EMB)*: Even faster broadband connectivity than 4G.
- *Massive machine connectivity*: The ability to connect many devices and often termed the internet of things (IoT).
- *Ultra-low latency*: Radio solutions that respond so quickly that immersive virtual reality (VR) experiences become seamless.

These themes are quite different and somewhat conflicting. For example, machines typically do not need high data rates but do need very long battery lives. The fifth generation is the first generation to have such widespread, different themes, and it may be that some deployments will concentrate on a particular theme.

Underlying these themes—a point that is central to the arguments made in this book—is that there must be a business case for moving to a new generation. For an MNO, there are costs associated with 5G, both for new equipment and for acquiring the new radio spectrum from the national regulator. This cost can easily run into billions of dollars. MNOs are shareholder businesses that aim to make a profit. Such an investment only makes sense if revenues grow or if costs can be lowered as a result. Investment in 3G was predicated on increased subscriber fees for mobile data usage (which did not really materialize). Fourth generation growth was predicated more on lower network cost through an optimized architecture leading to operational expenditure savings.

At the time of writing, the business case for 5G is unclear, with many MNOs admitting that they cannot yet make that case. The simplest argument for 5G at present is to add capacity to those parts of the network where the ever-increasing demand for mobile data is leading to congestion. With this deployment, 5G would simply be about more capacity rather than new features, and as a result, it would not be noticed by consumers.

Opinions Vary on 5G

A result of having multiple themes, as well as the lack of business clarity regarding the possible profits from converting to 5G, means that there are a range of views on what 5G is. Despite the fact that 5G radio spectrum has already been auctioned in some countries, that 5G has been trialed at events such as the 2018 Olympics, and that some MNOs are already claiming to have a 5G system, there is no consistent view around the industry as to the definition of 5G. This is an unusual and rather incredible situation and points to an underlying malaise that has not existed in previous generations. Put simply, it is not clear what the need is for 5G. Without a need, there will not be the increased revenue needed to fund its deployment.

In this book, when talking about 5G, the broad assumption is that it means at least one of the three elements described above—a better broadband experience than 4G, IoT connectivity, and low-latency communications. Insomuch as there was industry consensus, in 2018 it was considered that 5G would start with EMB, deploying in cities where there are capacity issues rather than being rolled out nationwide. However, there are important regional differences: for example, governmental pressure appears to be driving faster and more widespread 5G deployment in China; and competition may be driving some form of deployment for "bragging rights" in the United States, but in Europe the pressure appears to be less.

The net result is that it is not possible to give a clear and concise definition of 5G. Some have suggested that "5G will be whatever radio system happens to be deployed in 2020."

5G Technology

The introduction of new features typically requires new technology—if the technology is not there, then these features would have been deployed in the previous generation. For the first four generations, the new technology has concentrated on the radio part of the network, with new forms of modulating the signal, or radio waveforms, as the mechanism to enhance performance. The fourth generation also changed the core network, although this was more of a simplification than the introduction of anything new.

With 5G, despite much research on new radio waveforms, none have been found that make a material difference to performance. Instead, the focus has been on antennas. A 5G network makes use of beam forming, where antennas focus the radio energy like a torch rather than spread it around like a lightbulb. The term most often used for this is multiple input, multiple output (MIMO) antennas. However, MIMO is common in 4G. The difference in 5G is in the scale—that is, from 4G's nearly eight antenna elements at the base station to 5G's about 128. "Massive MIMO" is the term used to differentiate 5G MIMO from 4G.

To enhance flexibility, 5G also sees further changes in the core network, allowing the network to be "virtualized"—in that it can be deployed as a software solution on general-purpose computing platforms or cloud servers. This may bring cost savings to the MNOs and might enable new services to be introduced more quickly and flexibly than has been possible in the past.

Both of these technologies are challenging, and there is much yet to be learned before it will be clear how effective they are and what benefits they can really bring. The core network changes also risk network outages if the new platforms are not utterly reliable and so will probably be introduced with much caution.

5G Radio Spectrum

Each new generation has typically made use of additional radio spectrum, typically enabled by clearing older non-cellular applications from particular bands. Third generation initially concentrated on 2.1 GHz and 4G on 800 MHz, although there were regional differences and other bands were subsequently added. Just as 5G has multiple themes, it also has multiple bands:
- *A low band at 700 MHz*: This is anticipated to be used for widespread coverage, although broadly that already exists with 4G.
- *A mid-band at 3.5 GHz*: This will likely be the key band for 5G, used for EMB and additional capacity, although since it is at frequencies above those previously used for mobile, it has a reduced range.
- *A high band at 28 GHz*: This might be used for low latency or for other applications such as fixed-wireless access (FWA)—a way to provide home broadband without a cable or fiber connection.

The use of multiple bands further confuses the picture as to what 5G is or might be. In addition, the mid to high bands bring new problems such as low range that will require experience and research to fully resolve.

What the Consumer Might Expect

This chapter has shown that there is no easy answer to the question, "What is 5G?" At one extreme it might just mean additional capacity in cities, with consumers not noticing any difference (although without 5G, they would have noticed a steady degradation in performance as congestion grew). At the other extreme, it might be a swathe of new services enabling virtual reality, connected machines, and competitive home broadband provision. Opinions as to where in this spectrum 5G will finally land lie across the entire spectrum. One of the key purposes of this book is to analyze in a logical and evidence-based manner what the most likely outcome will be.

Chapter 2
Learning from Prior Generations

A Quick Trawl Through History

The reason that a fifth generation of wireless communications is being developed is because there have been four prior successful generations which have progressively taken cellular from a voice-only service to one delivering high speed data connectivity in three steps. Figure 2.1 shows the key trends moving through the generations.

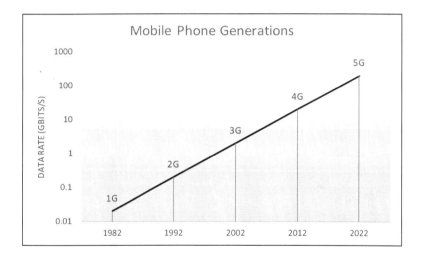

Figure 2.1: Trends in Mobile Phone Generations

The two aspects that are clear from Figure 2.1 are a regularity in timing and a steady improvement in data rates.

Generations have appeared every decade with great consistency. This may be predominantly because all the various steps needed from research, through standardization, to design and production take this long to develop. It may also be somewhat self-fulfilling as the industry anticipates a cycle of this length and so tends to work toward it for the next generation.

DOI 10.1515/9781547401185-002

Each generation has also resulted in a tenfold increase in data rates. This is somewhat more difficult to chart, as each generation has evolved during its decade, meaning that data rates often improved throughout its ten-year period. So, picking any one data rate for a particular generation is somewhat arbitrary. Also, the peak data rates quoted are rarely realized in practice. Figure 2.1 aims to select practical data rates. Hence the assumption of about 200 kilobits per second (Kbps) for the General Packet Radio Service (GPRS) element of 2G; about 2 Mbps for the High Speed Downlink Packet Access (HSDPA) evolution of 3G; and about 20 Mbps for the early deployments of 4G. Whether these are exactly right is of less relevance than the observation of the trend.

Figure 2.1 also plots 5G as an extrapolation. It shows 5G arriving in 2022 with practical peak data rates of around 200 Mbps. Many claims for 5G are well above this number, with claimed data rates in the gigabits per second (Gbps) range and with arrival dates set earlier than 2022. There are a significant number of claims of this sort each week. For example, *Business Insider* has said:

> 5G networks will have more advanced capabilities than their predecessors, according to the Next Generation Mobile Networks Alliance. These improvements will include capabilities like data load rates in excess of several tons of megabytes per second, enhanced coverage, and a significant reduction in latency—the amount of time between when data is sent from a connected device to when it arrives back to the same device.
>
> 5G will be instrumental in the next evolution of connected devices, including cars, smart homes, and wearables, due to its superior network speeds (10 times faster than 4G) and capacity (1,000 times the capacity of 4G). (Business Insider 2016)

Alternatively, Ericsson has published the following:

> There is a general industry consensus regarding the increase in demand that 5G systems will need to meet in comparison with today's networks. This common understanding indicates that traffic volumes will be multiplied 1,000 times; 100 times more devices will require connectivity; some applications will demand data rates 100 times the speeds that average networks currently deliver; some will require near-zero latency; and the entire system will work to enable battery lives of up to 10 years. (Ericsson 2015)

Many press releases focus on the data rates that can be achieved. Peter Dinham from iTWire wrote:

> Optus and Huawei have claimed a single user transmission speed of 35 gigabits per second was achieved over the 73 GHz band in a 5G speed trial just completed in Sydney. (Dinham 2016)

This, then, is the vision of 5G as portrayed by the industry: predominantly, about speeds that are 10 to 100 times faster than 4G and capacity levels that are 1,000 times greater.

Most books on 5G explain how this will be achieved. This book is different. It explains why the extrapolation from previous generations will no longer hold true and why 5G, as currently envisioned, will not be realized. In doing so, the book makes a number of observations:

1. Demand for higher speeds is unlikely to continue to grow.
2. Technological improvement will no longer be able to deliver more increased capacity without huge increases in cost.
3. Users will be unwilling to pay more for their mobile communications, limiting MNOs' appetite for investment.
4. Alternative futures—such as delivering consistent connectivity, which would be preferred by users and are more likely to improve productivity than the 5G vision—are more appealing.

Before moving to these observations, this chapter sets the scene with the history of transitions from previous generations and by looking at what drove them and the lessons learned.

Moving from 1G to 2G

The first generation of mobile phones was a revelation. It enabled those who were wealthy enough to make telephone calls without being tied to a landline. Phones were extremely bulky and expensive and reflected the difficulty of the engineering at the time. Networks were based on analog technology, where speech waveforms directly modified the radio bearer (rather than digital technology, where the waveforms were encoded onto digital bits first).

As 1G was increasingly adopted, it became clear there were many problems:
- Security was the most critical problem that allowed conversations to be readily eavesdropped by those with scanners, leading to scandals such as that involving Diana, Princess of Wales. It also allowed relatively easy cloning of phones, allowing hackers to steal identities and run up large bills. Toward the end of the 1980s, these issues were becoming extremely serious.
- Another problem was fragmentation, with different technologies in use in many countries. This prevented economies of scale as well as any form of roaming.

- The cost of delivering services was high with newer technology promising savings.
- Finally, quality and capacity of analog transmissions were relatively low, and evolving technology allowed better voice quality to be delivered using digital solutions.

The second generation resolved all of these issues to a fair degree. It added security that has not, even 30 years later, been breached in any material manner. It eventually provided a major improvement in voice quality through digital encoding,[1] and, through a mix of new spectrum and better reuse of frequencies,[2] it enabled a growth in capacity that allowed a network to support an order of magnitude more users. Fragmentation of technology still persisted but to a lesser degree. Europe consolidated on the Global System for Mobile Communication (GSM) while the United States and Japan had their own technologies. As a result, roaming was possible throughout Europe and to other countries in the world adopting the GSM standard but remained difficult within the United States.

The use of digital technology also allowed 2G to offer data transmission, although this was of lesser importance at the time. The unexpected and widespread use of the short message service (SMS; more widely known as texting) showed the community the potential of data use, leading to subsequent evolutions such as GPRS that enabled data capabilities to better match requirements. This was a highly successful evolution, delivering a materially better technology that overcame the problems of the previous generation and set the stage for the dramatic growth in adoption of cellular technology.

Moving from 2G to 3G

The 1990s saw the increasing adoption of the internet on desktop computers. Those involved in cellular technology predicted that the internet's benefits would be wanted in mobile phones but understood that 2G could not provide sufficiently high data rates to deliver an attractive service. There was also a view at the time that video calling would evolve from voice calling and that the 3G mobile network would need to support this. Hence, the key objectives of 3G were:
- To deliver higher data rates to enable internet browsing.

---

[1] Although some of the earlier voice codecs were relatively poor, these improved over time.
[2] This mix was accomplished through the use of time division multiple access that enabled better sharing among users, with less need for guard bands between them.

- To provide support for video telephony and the use of cameras to add pictures to texts (multimedia messaging service, or MMS).
- To enable both of these technologies by delivering greater spectrum efficiency, allowing for much higher data throughout.

The emergence of 3G happened at a time when a new access method had been pioneered by Qualcomm in the United States as one of their 2G solutions. Termed code division multiple access (CDMA), it promised significant improvements in spectrum efficiency by evenly spreading interference across all users. After much debate, it was decided to adopt CDMA as the underlying technology for 3G around the world.

Even with this history, effective implementation of the 3G version of CDMA proved difficult. Early 3G networks did not deliver high data rates and were difficult to plan and manage. Cell range in the frequency bands provided at 2 GHz was small, resulting in the need for many new base stations. Cells "breathed" as they loaded, reducing in size as more customers accessed them. The mix of circuit-switched and packet-switched traffic proved difficult to manage. Perhaps this did not matter that much, as mobile internet adoption was slow, with users finding it very difficult to browse on the small-sized screens of phones of the early 2000s. Also, video telephony did not prove as popular as envisioned, as cell phones had small screens, ill-placed cameras, and high per-minute costs.

Evolution of 3G slowly addressed these problems, with high speed packet access (HSPA) finally enabling and even exceeding the data rates originally promised. This moment coincided with the introduction of the iPhone, with its easy-to-use large screen and user interface that transformed internet browsing, causing extremely rapid growth in data demand.

Moving from 3G to 4G

Despite the improvements delivered by HSPA in its various forms, it was clear that 3G had not completely satisfied user performance demands. In particular, while various generations of HSPA radically improved the data rate, the latency of the technology (i.e., the delay between requesting information and receiving it) was still unacceptably long. The mix of circuit and packet switching made the networks less efficient and costlier to manage. The aim of 4G was to fix these problems. It did so with a different air interface, termed orthogonal frequency division multiplexing (OFDM), and by the removal of circuit switching. It also used wider frequency channels; this significantly improved the latency. The lack of circuit switching meant that voice calls could not be handled in the manner adopted in previous generations. Only some five years after this mix of technology was introduced is voice finally being carried over 4G using "voice over long-term evolution" (VoLTE).

The more stable and data-optimized networks offered by 4G meant that higher data rates could effectively be delivered to mobile users. Hence, the popular perception was that 4G was significantly faster than 3G. It was also around 2.5 times more spectrum efficient than 3G, allowing an important improvement in network capacity.

Some have noted that only the even-numbered generations have been truly successful, with 1G and 3G having significant flaws that were resolved by 2G and 4G, respectively. Whether 4G has flaws that need addressing through a fifth generation or whether 5G will suffer the curse of the odd-numbered generations is the topic of the rest of this book.

New Entrants into the Game

Over the last decade there has been a shift in the key manufacturers in the marketplace. The first four generations were led by long-standing large European and US companies such as Ericsson, Nokia, Alcatel, Qualcomm, Lucent, and Motorola. But during the 2000s many of these companies suffered, and in the early 2010s companies like Nokia, Alcatel, and Lucent merged or fragmented, becoming much weaker players with less research and development (R&D) capability and less budget to drive global standards development. In their place came Asian-Pacific companies, such as Huawei, Samsung, HTC, and LG. These new players initially adopted a tactic of being fast followers—letting others set the standards then delivering lower cost product within a few years. But as the 2010s progressed, these companies grew in stature and confidence; they became more significant global players in defining the role of new generations and delivering the research and standards needed.

As new entrants, the Asian-Pacific companies had to find a way into the "system" and tended to focus on showing that they could deliver faster and better technologies. This has changed the dynamics of the industry, with manufacturers racing to deliver 5G sooner and with faster data rates than their competitors. The implications of this will be considered in subsequent chapters.

Extrapolations Do Come to an End

By way of comparison, consider being in the year 1970 and trying to make predictions about airline speed. A plot of the speed of key airplanes against time is shown in Figure 2.2, along with a trend line.

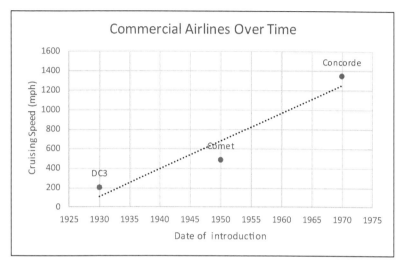

Figure 2.2: Airline Speed in 1970

The extrapolation would be clear—speed increases by about 600 mph every 20 years. But now consider what actually transpired, as shown in Figure 2.3.

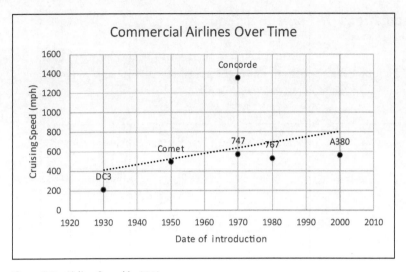

Figure 2.3: Airline Speed in 2010

Concorde was an outlier. After its introduction, speeds settled back to broadly those of 1950. Had companies or governments embarked on the design of a 2,000 mph airliner in 1970, they would have gotten it badly wrong. The reason for the breakdown in the extrapolation was economic rather than technical—designs for 2,000 mph airlines do exist, but they would be much too expensive to operate. Often it is economics, rather than technology, that causes these "laws" to come to an end. Perhaps the same is true of mobile speeds in 2016. The introduction of 5G could end up being like Concorde—a superb feat of engineering but of limited value to all but a small minority. Whether there is any evidence that this might be the case is the subject of Chapter 3.

Chapter 2 References
Business Insider. 2016. "Huawei Has Taken the Next Step in Setting the Standard for 5G." *Business Insider, December 1.* https://www.businessinsider.com/huawei-will-lead-the-way-for-5g-standard-2016-12.

Dinham, Peter. 2016. "Optus, Huawei Claim Speed of 35Gps with 5G Trial." *iTWire, November 16.* https://www.itwire.com/telecoms-and-nbn/75762-optus,-huawei-claim-speed-of-35gbs-with-5g-trial.html.

Ericsson. *5G Systems: Enabling Industry and Society Transformation.* January 2015. https://www.ericsson.com/assets/local/news/2015/1/what-is-a-5g-system.pdf.

Chapter 3
Can Demand Grow Indefinitely?

Volume Versus Speed

Demand for mobile services is typically measured in monthly data volumes—a metric of the quantity consumed. However, the discussion so far has suggested that the key reason for introducing new generations of mobile technology is speed. This chapter discusses how capacity and speed are related before considering the likely growth in demand for both speed and data in 5G deployment.

A simplistic approach might assume that there is no relationship between speed and data. For someone who wishes to download an attachment or a video, the volume of data is unchanged regardless of speed—it just takes longer on a slower connection. In practice, there is some correlation. Users will only attempt certain activities, such as streaming video, when the data rate is high enough. Hence, once they pass certain thresholds, higher data rates tend to trigger step-changes in demand. MNOs have observed this effect by noticing that data demand tends to rise significantly as subscribers who have moved from 3G to 4G get faster and more reliable data. However, it might be expected that, once speeds are available beyond those needed for data-intensive applications, the link between speed and demand would again fade away.

New generations of mobile technology typically aim both to:
1. Provide higher data rates in order to improve the user experience.
2. Provide more efficiency in terms of throughput per unit of radio spectrum to allow networks to carry increased traffic.

This chapter considers the extent to which this is needed in 5G.

What Speed Is Needed

The data rate that is considered acceptable has progressively grown over time. As higher speeds have become possible, application providers have delivered new services, driving the demand for these types of speeds. For example, once speeds rose above 1 Mbps, delivery of video content became possible, spurring a huge growth in demand. This then led to the delivery of higher resolution video services, which again spurred demand.

Determining how much speed is "enough" is problematic. The speed that is sufficient at a given point in time for the services typically consumed can be calculated. New services may be devised that result in a greater demand for speed, although this is tempered by the fact that apps are increasingly being developed for a global market. China and India are far larger markets than Europe but have lower speeds of connectivity. Hence, even if very high speeds are provided in developed countries, the focus of app developers may nonetheless be on lower speeds to access larger markets (techUK 2016).

Demand for the highest speeds and the largest data volumes is almost invariably driven by video consumption. A person can only watch one video stream at a time, so understanding the data rates associated with the highest quality video feed required is a good upper limit. This may increase in the future if higher resolution video grows (e.g., demand for 4K video) or if applications such as virtual reality demand more information. By way of example, 4K video (video with roughly 4,000 pixels of horizontal resolution) requires around 20 Mbps.[1] This is an upper limit, as most mobile screens are far too small to make watching video at this resolution worthwhile. MNOs have found that "throttling" video back to 1 Mbps or even less has no noticeable impact on mobile handset users.

A somewhat different question is the speed needed for instantaneous web browsing. The issue here is less about absolute speed and more about "latency"—the time it takes for a request (e.g., for a new page) to be sent to a server and to receive a response (as shown in Figure 3.1). Beyond a certain speed, other factors—such as the turnaround time at the server and the delays inherent in the internet TCP/IP protocols—become constraining.[2] This data rate is currently around 4–8 Mbps (and hence most users will not notice an improved browsing experience once data rates rise above this point). Resolving this requires changes to internet protocols and architectures—something that has to occur on an international basis within internet standards bodies and key industrial players.

---

[1] For example, see Netflix recommended data rates (Netflix n.d.). Here they suggest 5 Mbps for high definition (HD) and 25 Mbps for 4K ultrahigh definition (UHD). Others expect coding efficiencies to improve and the UHD rate to fall (for example, see Rosenthal 2014).

[2] For a detailed discussion, see Getty (2013). Also, see Figure 5 in Sundaresan et al. (2013), which shows the same effect.

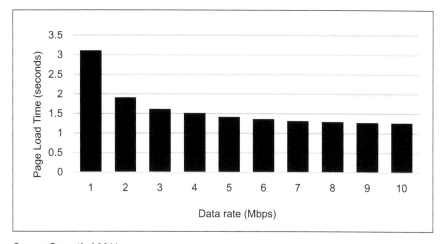

Source: Rosenthal 2014.
Figure 3.1: Page Load Times

There is also something of the-chicken-and-the-egg issue in that, if data rates reach a point where the costs of increasing them further are very high, then developers will be aware of this and will not tend to develop applications that exceed this data rate. Hence, the point where a major technical or economic step-change is needed to increase data rates can become "sufficient" in a self-fulfilling manner.

Alongside mobile coverage provided via cellular, there is also mobile coverage provided via Wi-Fi. Indeed, estimates are that of all the data provided to mobile devices, about 50 percent to 80 percent[3] is carried via Wi-Fi. Wi-Fi also typically provides the connectivity in the home from the broadband access point. Wi-Fi can, in the best-case scenario, deliver rates in excess of 50 Mbps. However, these rates drop quickly if the Wi-Fi spectrum is congested or if devices are not close to the access point.

The sufficiency of the currently available speed was confirmed in a recent study by the BCG [3] (Sherman et al. 2015) which asked users about their mobile experience and correlated it with the network speed and latency. They found that once speeds exceeded 1.5Mbits/s there was no increase in consumer satisfaction when watching video, and that satisfaction when using apps stopped increasing at lower speeds, below 1Mbits/s. They concluded that "the telcos' race for

---

[3] Mobile coverage via Wi-Fi is very difficult to quantify as most Wi-Fi access points are self-owned and the traffic carried through these generated by mobile phones is not counted. The best approach is to run an application on the handset that monitors how data is sent, but this requires user permission.

speed may, in fact, be a largely unnecessary endeavor that breaks the cardinal rule of focused investment: spend where the spending counts most." They did notice that satisfaction increased as latency fell, with latency levels as low as 75ms appreciated for video and in the region of 25–50ms for some other apps (4G currently delivers around 30ms latency in most uncongested networks). They concluded that "cellular carriers should, therefore, consider intensifying their efforts to reduce network congestion."

As BCG noted, current 4G networks deliver speeds well in excess of 10 Mbps (when not congested). Indeed, some MNOs now market speeds of above 100 Mbps, and manufacturers have reported tests delivering over 1 Gbps when aggregating multiple 4G carriers. *It is very hard, then, to understand why speeds faster than 4G might be needed.* (The question of whether more capacity is needed in order to ensure that speeds do not slow as demand rises is considered next.)

In passing, it is worth noting that there are parallels here with the fixed broadband world. The same restrictions apply on web browsing and broadly the same on video download. For home usage, the highest data rate service is likely to be 4K streaming video using around 20 Mbps. With multiple occupants in the home, requirements might peak at around 60–80 Mbps. This is well within the capabilities of fiber-to-the-cabinet (FTTC) deployments which then use solutions such as very high bit rate digital subscriber line (VDSL) or G.fast to the building. The focus on delivering fiber to the home seems as equally unnecessary as the high speeds within 5G. Further, it might cannibalize investment that could otherwise occur in important areas such as enhancing coverage, as examined in later chapters. Indeed, in Australia where there has been government-sponsored deployment of fiber to the home (FTTH), the experience has been that (1) few subscribers opt for the higher data rates available via fiber, with most selecting rates of 25 Mbps or 50 Mbps that would have been available on FTTC, and (2) the additional time and manpower needed for FTTH deployment has meant that homes that would have already received an FTTC upgrade under an FTTC strategy are still awaiting any form of upgrade. Here, the net result for the consumer has been negative.[4]

Latency

As noted above, latency does have an impact on user experience. Latency is broadly a measure of how long it takes for a request to make it from the mobile phone to the end point in the network. For applications like video streaming,

---

4 See, for example, Williamson 2017.

latency is irrelevant, as the handset can buffer material in advance and so never needs to send an urgent request. But for applications like web browsing, once speed is above 1 Mbps, it is latency that determines the time it takes for a new page to load after the user clicks on a link.

The importance of latency can be seen when web pages have numerous elements. It is quite common for a web page to have 100 or so parts—text, images, headers, advertisements, and so on. If the latency was 30 milliseconds (ms) and there were 100 elements all fetched sequentially, then the page load would take 3 seconds. In practice, some elements can be fetched in parallel; web pages optimized for mobile use have much fewer elements.

Latency has fallen across the cellular generations. It was around 500 ms with 2G, perhaps 100 ms with 3G, and around 30–50 ms with 4G. As BCG noted, falling from 100 ms in 3G to 50 ms in 4G has improved user satisfaction.

Further improvements may be both harder to deliver and have less impact. As an example, most applications where latency is seen as a critical issue involve video—such as VR or remote control applications. Even the most advanced VR headsets have a video refresh rate no greater than 100 hertz (Hz), which means it takes 10 ms for a new video frame to appear. The frame requires some processing before it can be displayed. Hence, even with zero latency from the rest of the system, around 15 ms latency is inevitable.

The theoretical latency of 4G is 10 ms—additional delays tend to occur within the core network of the MNO. Once the message leaves the cellular network it may need to traverse continents. The latency imposed by a message sent from the East Coast of the United States that needs to reach a server on the West Coast is around 30 ms, and the time for it to get from Europe to the United States is around 60–100 ms. Some content can be cached on servers within the same country, but this typically only works for the most visited pages. So, a 4G message might incur a 10 ms delay across the radio interface, 30 ms across the MNO's core network, and 50 ms across the internet. In this case, the radio delay is slightly over 10 percent of the total. If the radio delay were reduced by half to 5 ms, the total delay would fall from 90 ms to 85 ms, which would be barely noticeable to any users.

Of course, core networks could be optimized to reduce the 30 ms delay, but this can be done with 4G networks perhaps more easily than a complex virtualized 5G network, where many of the resources in the core would not be under the control of the MNO.

Currently, work is underway to reduce the theoretical 4G latency to 5 ms by halving the size of resource blocks. Within the 5G community, the original 1 ms latency target had been determined impractically difficult and modified to 8 ms, although these numbers may change as standards develop.

Latency is important, but 5G seems unlikely to have a materially lower latency than 4G in practical situations. Even if 5G did have a lower latency, the impact would be minimal, since latency in other parts of the networks would dominate.

Predictions of Volume

The volume of mobile data has been growing ever since the launch of the iPhone in April 2007. Figure 3.2 shows what happened to both voice and data volumes in the years immediately after the launch of the iPhone.

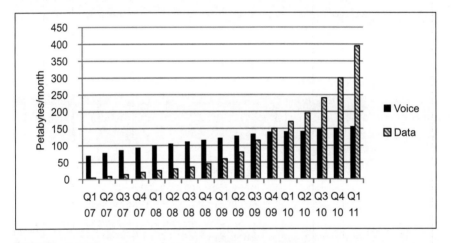

Figure 3.2: Growth in Mobile Data After the iPhone Was Launched

While voice grew steadily, data demand exploded, growing around a hundredfold in the five years from 2007 to 2011. It has continued to grow rapidly since then. Figure 3.3 shows the predictions for mobile data traffic made in Cisco's industry renowned "virtual networking index" (VNI) in recent years.[5]

---

5 The Ericsson Mobility Report (Ericsson 2018) comes up with similar figures.

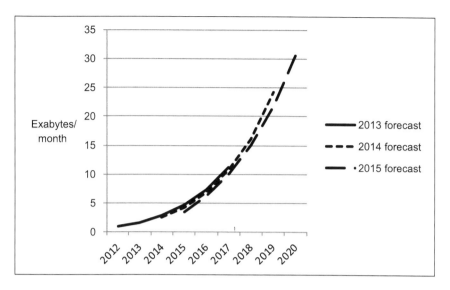

Figure 3.3: Cisco's VNI Predictions of Mobile Data Traffic Growth

The figure shows that demand continued to grow and is predicted to do so for the foreseeable future. However, there are some interesting points to note. The forecasts have been trending slightly lower. For example, predictions for 2017 that were made in 2013 were 11.2 exabytes per month; in 2014, they were 10.7 exabytes per month; and in 2015 they were 9.9 exabytes per month. It's important to note that growth is slowing. This is shown in Figure 3.4, where the predicted growth rate from one year to the next—along with a linear trendline—is shown.

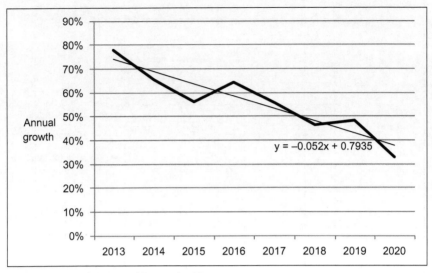

Figure 3.4: Growth Rates and Regression Line

Data growth is predicted to fall to 37 percent per year by 2020 and, if the trend continues, to zero by 2027. This is only five years after the likely introduction of 5G, assuming the "one generation per decade" trend holds. If this were to occur, then the mobile data requirements over the period 2007–2027 would be as shown in Figure 3.5.

This figure shows a classic "S-curve," which is experienced frequently with many new services and devices. If this were to occur, data would plateau at around 10 times today's levels, or around 15–20 gigabyte per user per month. Such a plateau might be expected. There is only so much data that a mobile subscriber can consume. Once they are watching video in their free moments while downloading updates and attachments, there is little more that they could usefully download. At the level of the plateau shown here, the average mobile user would be consuming over an hour of video via cellular on their mobile device every day. (Note that there may also be video streaming to the mobile device via Wi-Fi, and video streaming to other devices such as tablets also likely via Wi-Fi, so this represents video mostly consumed on the move.) Of course, some users will exceed this, but equally many will consume much less or will download video in advance. Personal data usage must plateau at some point.

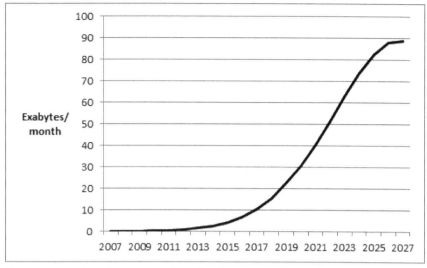

Figure 3.5: Data Growth, 2007–2015 (actual) and 2016–2027 (forecast)

There is evidence that this is already happening. For example, the Singaporean operator M1 reported almost flat data usage over the year 2015–2016:

> Average postpaid smartphone data usage was 3.4GB per month in third quarter of 2016, from 3.3GB per month a year ago. Mobile data revenue increased 6.2% percentage points year-on-year to 54.2% of service revenue in the latest quarter. (M1 2016)

Similarly, Tefficient reported:

> At the same time Singapore, Sweden, Japan, Hong Kong and Portugal show significantly slower usage growth—in Singapore's case just 24%. (Tefficient 2016)

The Tefficient report looks for areas of correlation between growth and other variables and concludes that the key is price and that users pay around €20 per month, consuming as much data as they can for this price. Countries with high data usage are those where the cost per byte is relatively low. This is a point returned to in Chapter 5.

A paper from LSTelcom (Realwire 2017) looked at the maximum possible demand that users might want, assuming consumption of 4K video for the whole 16 hours per day that people are awake, and demonstrated that even this extreme level implied a plateau of demand. It then studied a number of countries where demand was already slowing and showed that growth in Singapore, Japan, and Sweden had fallen to between 10 percent to 20 percent per year. The relevant curves for Singapore and Sweden are shown in Figure 3.6 and Figure 3.7.

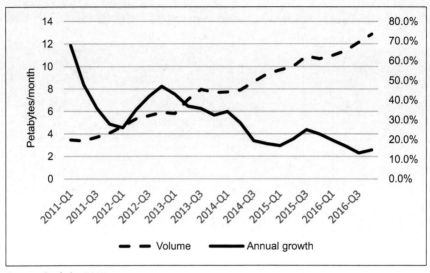

Source: Realwire 2017.

Figure 3.6: Data Growth in Singapore

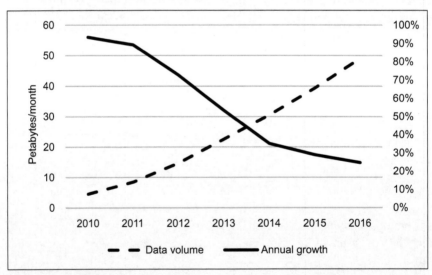

Source: Realwire 2017.

Figure 3.7: Data Growth in Sweden

It is not completely obvious why these countries might have lower growth than average. Since they already have high usage levels, it is possible that they are among the first to reach the plateau.[6] Some have speculated that they may have an indoor culture and good Wi-Fi provision, resulting in more substantial off-load to Wi-Fi than elsewhere. This sounds plausible, and if so it might be expected that other countries would also progressively improve their Wi-Fi. It seems unlikely that their general usage patterns of apps and video would be materially different. Broadly, then, these may be the first signs of a slowing in growth that will occur globally over the coming years.

In late 2018, Verto Analytics published a report covered by the BBC (Marston 2017) which showed that, for the first time ever, the amount of time spent on a mobile phone for people aged 16–24 had fallen. Interestingly, growth in usage is now occurring in the older generation, who are slowly catching up with their children and grandchildren. This also implies that a plateau in usage will occur and has already been reached by the heaviest users. As others increasingly "catch up," overall data usage growth will fall.

All of these indications suggest that the predictions in Figure 3.5 might actually be too high and that growth rates might fall more quickly than suggested there. Certainly, industry expectations of a 50 percent growth per year ad infinitum look hugely optimistic.

Why then does the industry—from manufacturers to operators and from regulators to governments—continue to express concern about data growth? And why do 5G advocates call for a data capacity growth of 100 times or even 1,000 times as compared to the level in 2015? Perhaps this is because growth has been exponential since 2007, with the result that many companies were caught by surprise. Regulators have struggled to find sufficient spectrum and MNOs to deploy the needed capacity. With such memories being very recent, it is natural to assume that growth will continue to surprise. It is also safer for those who do not have to pay for increased capacity to overpredict rather than to underpredict. Few would question a regulator that forecast fast growth. A forecast of slowing growth might be criticized as lacking ambition or subverting the activities of the industry. It's safer to over forecast. And if all companies are doing it, there is a self-reinforcing safety in numbers that drives ever-higher predictions.

Of course, there is the possibility of another "iPhone moment"—the emergence of a device or service that causes another step-change in demand. Perhaps this might be connected IoT devices (or something similar). This is discussed below and in Chapter 7.

---

[6] There are a few countries with much higher usage levels—notably, Finland. This appears to be driven by a significant use of cellular as a home broadband substitute and so may not reflect true mobile usage.

Interestingly, when considering the feedback from the first edition of this book, the vast majority was positive and agreed with book's premise. However, the one area where opinions diverged was in the view that data growth would plateau. Here, some felt that there would be some other application, as yet unidentified, that would spur further data growth in due course. Of course, they may be right; but the issues with the "build it and they will come" view of the future are many and are discussed in more detail below. An alternative view is that, having been caught unawares by a data growth rate that was faster than expected, many analysts are now compensating by over forecasting.

If Figure 3.5 is correct and if 5G is introduced in 2022, then total data growth over its lifetime will be around twofold, likely matched by additional spectrum—meaning improved spectrum efficiency is not needed. Indeed, given that it will take some time for 5G devices to become widespread[7] within the population, it is quite likely that growth will be near zero by the time 5G technology makes any effective difference to network capacity.

Another way of considering this is to ask how MNOs can deliver 10 times more capacity over the next decade. The position of each MNO differs, as it depends on the MNO's spectrum holdings, current subscriber usage, and base station density; so broadly speaking, this growth can be achieved by the following:

1. Refarm 2G and 3G spectrum holdings to 4G. For many MNOs this will provide a two- to fourfold improvement in capacity over their current networks.
2. Add more spectrum. The addition of spectrum that is likely to be released in the next few years at various bands around the world[8] includes 600 MHz, 700 MHz, 2.3 GHz, and 3.4 to 4.2 GHz. This may add a two- to threefold improvement, especially as much of the new spectrum can be used for time division duplex (TDD) where it can be biased toward downlink capacity and where MIMO antenna solutions can work most effectively.
3. Additional gains. Any remaining gains might arise from improved MIMO operation (discussed in Chapter 4), more off-loading to Wi-Fi, and perhaps the use of unlicensed spectrum in some cases.

These improvements will broadly provide the capacity predicted to be in demand. If they fall slightly short, then increasing prices for larger data bundles will likely

---

[7] Recall that the GSMA predicted only 40 percent of the population would have 5G coverage by 2025, and this is likely to be an optimistic forecast given the GSMA's membership.

[8] Some of these have already been released in some countries; for example, 700 MHz and 2.3 GHz is already available in the United States.

enable a match of supply and demand at a somewhat reduced level from that predicted here.

One other trend worth noting is that handset replacement cycles are gradually lengthening. For example, in 2015 it was noted:

> Still, the latest smartphones are not being replaced as quickly as they were three to five years ago. Apple has seen its replacement cycle of iPhones increase from 21.7 months two years ago to 27.4 months this year, according to Roger Entner, an analyst at Recon Analytics. (Hamblen 2015)

This trend continues as manufacturers struggle to find new features that make a cell phone upgrade compelling. This means that even if MNOs introduce new technology such as 5G in their network, it may be many years before a substantial proportion of their subscriber base has devices that can access the new network. This increases the length of time that 4G would need to carry traffic and also the time needed to achieve a return on the investment.

One question is whether there might be new applications that result in a new wave of traffic not included in the forecast above. The key possibility mentioned is the IoT.

Internet of Things

The IoT may have the potential to generate substantial new volumes of mobile traffic. In this section, the IoT is discussed and the argument is presented that the volumes of data generated are likely to be very small. (Note that autonomous cars, drones, and robots will be covered in Chapter 7.)

What Is the World of Connected Machines?

There is much discussion about the world of connected machines. This is sometimes called machine-to-machine (M2M) or the IoT. Most agree that there is huge potential in connecting machines, but this marketplace is vast and diverse. There is much confusion as to what "connected machines" entails and how it will materialize.

A first step is to differentiate different areas of the marketplace. One key differentiator is whether the system is really just a remote control for a machine (e.g., being able to regulate one's home heating system from a smartphone app when outside the home), or whether it is divorced from any person and is a machine reporting on status (e.g., a dustbin notifying a database that it is full). The "remote control" market is typically handled through cellular, Bluetooth, and Wi-Fi and is already established and growing quickly as seen at key consumer and trade conferences such as the Mobile World Congress. The data volumes this will generate are already included in cellular data forecasts.

Another key differentiator is whether the machine communication is within a building, such as the home or office, or whether a "wide area" communication is needed. A home system might control the lighting in the building. A wide area system might send smart meter readings to an electricity supplier. Broadly, the former is "home automation" and can be delivered effectively using solutions such as Bluetooth and Zigbee. Again, this is not discussed here.

This leaves us with wide area, machine-oriented solutions. These solutions form the bulk of the interesting new areas that people discuss regarding M2M and the IoT. Unfortunately, neither of these terms is accurate. These are machines sending readings to a central database or computer that can then take an action—such as sending an electricity bill or scheduling a garbage truck. This is not really "machine to machine" but rather sensor to database. This may seem to be pedantic, but there is an important point here: there is little reason for one "machine"—for example, a device such as a smart meter—to talk directly to another. Equally, it is not really an IoT. The internet implies an interconnected network where one computer can access information from another computer. Instead, only the "owner" of the machine, such as the electricity company, will be able to access its readings and communicate with the device. Consumers that wish to read this data will then retrieve it from a cloud-based server and not directly from the machine. It is more like the "intranet of things," where connectivity is restricted to self-contained groups rather than the internet. Again, this may seem pedantic, but it has important implications for installation, security, and network architecture.

Most envisioned applications are self-contained in that the data generated is used just for that application. For example, in smart metering, the meters report to electricity companies; with parking, the car park sensors talk to parking applications. There may be some cases where sharing data more widely than indicated here is valuable, but at present it looks like these will be exceptions. Most machines will only send information to one place—the client database—and will only receive information from the same source.

A connected machine in this sense is a device that communicates over a wide area network to its owner's computer system. This enables the computer to take appropriate action, such as replanning schedules. Boring, but incredibly valuable.

The Current Position

It is hard to state exactly where we are on the connected machines evolution. Ericsson, for example, predicts around 50 billion connected devices by around 2025. The size of the market seems entirely plausible—it would only be 10 devices per cell phone owner worldwide. To date there are a few hundreds of millions of devices deployed—perhaps 1 percent or so on the way to the vision of a connected world. Where connected devices have materialized, they tend to be either short-range devices in the home or high-value wide area devices. This is because short-range devices can be accommodated today with in-home network solutions such as Bluetooth and Wi-Fi, while high-value wide area devices can tolerate the cost and battery consumption of using cellular communications. Examples of the former are home security systems and sensors to remind us to water the plants. Examples of the latter are vending machines and high-end cars.

Most of the pieces needed for machine connectivity are now in place. Software that can gather data from machines across multiple networks and present it to the client's database system is now available from companies such as Jasper and Interdigital. This often includes intelligence that can simplify the deployment process, such that when a device like a smart meter is turned on for the first time it is automatically configured by the network software rather than requiring the installation engineer to enter details into some remote terminal. Systems such as smart meters are being rolled out in many countries. However, a key missing piece of the puzzle is a wide area wireless solution that provides very low cost ($2 hardware, $2 per year connectivity), a 10-year battery life, and ubiquitous coverage. Experience with concepts such as mobile data has shown that the market tends to bump along the bottom until all the pieces of the puzzle are firmly in place, at which point it only takes a small stimulus (the iPhone in this example) to result in explosive growth. If it were possible for a device manufacturer to buy, for $2 or less, a wireless module that they could simply add to their machine with the knowledge that it would work anywhere in the world with no need for complex roaming agreements, then this would likely stimulate the widespread deployment of connected devices. Of course, it takes some time for certain machines to be replaced or updated, so growth may not be quite as fast as in the smartphone arena but could be many orders of magnitude faster than it is at present.

If the wireless connectivity part of the puzzle is solved with appropriate technology and with vision from operators, all the other pieces needed may be in place for rapid growth.

Which Applications Will Lead?

There is understandable interest in predicting which machine applications will lead the connected machine world. Of course, as mentioned above, there are already some deployed applications, so it is not so much which machine applications will be first as which will take us toward the first one billion connected devices. Trying to predict what will happen next is just as difficult as predicting the top selling applications in the Apple Apps Store in advance of the launch of the iPhone. It may well be that multiple applications develop simultaneously, as has happened in the world of mobile data. The best that can be done is to make some general observations about some of the more obvious areas:

- Smart metering is clearly a lead application because of regulatory mandates in some countries that require rollout to happen over the coming years.
- The energy market (refineries, wind turbines, etc.) is also a fertile area for machine connectivity, albeit much smaller than the smart meter market.
- Smart city applications such as smart dustbins, rental bike monitoring, and street light management are promising because of the limited coverage requirements and the clear need for productivity improvements in delivering citywide services. However, this is balanced by the complex procurement and ownership issues where budgets may be held by local authorities but services subcontracted to commercial providers.
- The healthcare market is potentially enormous and valuable. Simply speaking, it divides into those solutions that can be directly purchased by consumers or private healthcare providers and those that require government approval or funding. The former might include enhanced pregnancy testers, pill dispensers with reminder functions, heart rate monitors, and fall alarms—all potentially linked to a home health hub. These could be quickly purchased and deployed. The latter includes automatic medicine dispensers and monitoring of key indicators such as blood sugar with alert functions. Because these items will often be regulated and because they may require linking to national systems, they will take much longer to gain approval.
- Consumer devices are a promising, if broad, category which might include devices like washing machines, connected TVs, or even home weather stations. Some of these will use the home network, but there are strong advantages in others having simple and direct connectivity back to the manufacturer.
- The automotive industry, as well as the railroad and airline industry, tend to be a late adopter of wireless technologies.
- Asset tracking is another potentially enormous market, but to realize its full potential requires widespread or even global coverage. For some applications this can be achieved with cellular networks, but global coverage is currently

lacking in cases where lower power consumption is required and necessitates an IoT-specific technology.
- Users such as the military might have a great demand for machine connectivity in areas such as tracking soldiers or key items of equipment or monitoring the status of weapons, but it is unclear as to how quickly they might deploy such solutions (or indeed if they already are).

Privacy and Security

Any system that gathers information and takes action on it will need to address legitimate concerns over privacy and security. Of these, security is the simplest and can broadly be addressed using appropriate encryption and authentication within the various layers of the system. It is helped by keeping communication between a machine and its client computer, thus preventing others "hacking in" to machines. Privacy is more complex. In some cases, there may be minimal issues—for example, few will be concerned about the privacy implications of whether their local street lamp is reporting a broken bulb or not. Other applications such as healthcare may raise deep concerns. Privacy will need to be addressed on an application-by-application basis and should demonstrate that the application delivers benefits to the end user that strongly outweigh any potential privacy issues.

Where Might It End Up?

What does a world of connected machines look like? Of course, it is almost impossible to predict. Connecting machines will lead to changed behaviors that then change the applications and so on. Equally, it is easy to get carried away with visions of an "internet" of connected machines. Broadly, connected machines will just work better. Washing machines will be better optimized for water hardness, can have new programs downloaded as new washing detergents come onto the market, and so on. Trash containers will be emptied when they are nearly full, and not on a routine weekly cycle. Meters will be read automatically. Cars will have software updates delivered wirelessly and so will not need to be recalled so often. Finding an empty parking space will become simpler. Broken street lamps will be repaired more quickly. Cracks will be spotted earlier in bridges and will be repaired with fewer failures or disruptions. Home devices will automatically connect to the correct home network, with no need to enter the password into the fridge. And so on. The world will be a better, less frustrating place to live in. Productivity will improve through less human intervention, thereby leading to growth. Some key societal problems such as assisted living will be ameliorated through sensors in the home and on the person. But in the same way that we only notice street lamps when they don't work, we may hardly notice that we live in a world of connected machines. Indeed, that could be the ultimate aim of the connected world—we stop noticing the machines at all.

Implications for Network Load

If the "50 billion devices" prediction is realized, each person with a smartphone will have about 10 devices. Of these 10, many will be connected in the home or in the office using Bluetooth or Wi-Fi. Others may be connected over proprietary IoT networks such as those deployed by Sigfox and LoRa or through emerging standards such as Weightless. A subset will be deployed on cellular networks. This might be perhaps three or four devices per person.

A typical device sends periodic readings, such as temperature or meter levels. Perhaps it might send information hourly, sending around 200 bytes per transmission. This would equate to 4.8 kilobytes (KB) per day or 144 KB per month. With four devices, that would be around 0.5 megabytes (MB) per month. The average mobile user currently consumes around 1.5 gigabytes (GB) per month. This implies that the IoT device load is about 0.03 percent of the current load on cellular networks. Even if these estimations are off by a factor of 100, the amount would only be 3 percent of current cellular load and 0.3 percent of the predicted 2025 load. Hence, it would seem unlikely that IoT traffic would make any material difference to overall load.

This could change if there were applications such as connected cars that had constant high-bandwidth connectivity. But it is hard to see why this would be required and hard to imagine why anyone would pay the necessary data transmission fees.

These types of IoT are very unlikely to materially add to data traffic volumes.

Could Anything Make a Difference?

The forecast suggested here is equivalent in 2027 to one hour of video via smartphone per person per day. For there to be a material difference in the forecast—that is, enabling growth to continue at 30 percent to 50 percent annually throughout the 5G era—the traffic would need to increase by at least 5 times, if not 10 times, more than this. This means, effectively, moving toward 5 to 10 hours of video per day sent via cellular. Even if autonomous vehicles (considered in Chapter 7) streamed video—which seems unlikely—the average person spends an hour a day in the car, delivering only a one-time gain. Around 30,000 IoT applications per person would be needed for another one-time gain. Broadly, machines do not communicate via video, and so machine-related traffic is unlikely to make a material difference. Changing to higher video streaming rates would be material, but these rates will reduce over time as codecs improve, and consumers will tend toward the rate that provides adequate quality at the lowest price rather than consume 4K video on a handset where the improvements are likely to be minimal. Applications such as body-worn videos (body cams) that continuously stream to cellular networks could make a difference, but it seems unlikely many users will pay the costs of such streaming when relevant sections of video can be stored and uploaded via Wi-Fi as needed. Augmented reality is not a large consumer of bandwidth—Pokémon Go, the much-played phenomenon of 2016, only added 0.1 percent to existing network traffic (Statt 2016).

The risk is that there are "unknown unknowns"—applications that emerge that we could not envision. However, no applications that materially change the bandwidth requirements predicted for 2020 onward have emerged unanticipated in the last 20 years. The key bandwidth driver is video, and this was foreseen well before 3G arrived.

There is a further argument that "if we don't build it they won't come"—the view that applications can only emerge once the underlying network is capable of supporting them. In this view, unless 5G networks capable of delivering superfast data rates and ultralow latencies are deployed, new applications will not emerge. In practice, applications often emerge on suboptimal networks and demonstrate potential; then network enhancement can be justified. This is typically a necessary step, as justifying investment in expensive new infrastructure on the basis of unknowable technology is challenging. The need to build a network before applications appear might apply in the case of the tactile internet (discussed further in Chapter 7), but even here it seems likely that many applications such as VR could be trialed indoors initially, with outdoor capability being provided if needed in due course.

In the GSMA's *Global Mobile Trends 2017* report (GSMA 2017), slide 31 is titled "If you build it they don't always come" and shows that in some countries, such as Germany, the usage of 4G networks lags well behind coverage. While "build it and they will come" may have worked for the data explosion in 3G, it has not been the case for all 4G operators. Clearly this casts doubt on whether building a 5G network is likely to achieve economic returns.

In summary:
1. No applications that subscribers are likely to be willing to pay significantly more for and that will generate five times the predicted 2027 traffic loads are currently foreseen.
2. No applications have emerged in the last 20 years that were both completely unexpected and resulted in substantial increases (e.g., 5 times) in total traffic volumes.[9]
3. Users are unwilling to pay more for their communications (see Chapter 5), so applications would need to deliver clear value.

On this basis, the probability would seem to be with the forecast presented here rather than one of continued growth throughout the 5G era (assumed to be the years 2022–2032).

This is not in any way to be pessimistic about the future—far from it. Within this growth forecast a huge array of new applications, connected machines, social networking, augmented reality, and much more can and will be delivered. New applications will abound, and mobile connectivity will enrich lives and lift productivity. Happily, this can all be achieved today with the technology and networks already at our disposal.

Conclusions

This chapter suggests that current mobile data speeds are more than adequate for all foreseeable uses. It also suggests that data growth is slowing and may plateau around 2027, with only about two times the growth during the 5G era. With 5G predicated predominantly on higher speeds and also on its ability to deliver substantially enhanced data capabilities, this suggests that the technology may not be targeting the right areas.

The next chapter looks at why it would be very difficult to deliver 5G technology, even if the 5G vision were correct in its assumptions as to speed and data volume requirements.

---

[9] Social media—mostly Facebook—comes the closest but only generates some 15 to 25 percent of today's network loading. If it were to disappear, it would have only a minor impact on increasingly video-dominated networks.

Chapter 3 References

Ericsson. 2018. *Ericsson Mobility Report Q2 2018 Update*. Stockholm, Sweden: August. https://www.ericsson.com/mobility-report?utm_source=Press%20release&utm_medium=earned&utm_campaign=Mobility%20Report.

Getty, Jim. 2013. "Traditional AQM Is Not Enough!" *jg's Ramblings*, October 24. https://gettys.wordpress.com/2013/07/10/low-latency-requires-smart-queuing-traditional-aqm-is-not-enough/.

GSMA. 2017. *Global Mobile Trends 2017*. GSMA Intelligence, September. https://www.gsma.com/globalmobiletrends/.

Hamblen, Matt. 2015. "Consumers Are Keeping Smartphones, Tablets and PCs Longer." *Computerworld*, September 24. http://www.computerworld.com/article/2985483/smartphones/consumers-are-keeping-smartphones-tablets-and-pcs-longer.html.

M1. 2016. "Results for Nine Months Ended 30 September 2016." News release, October 18. https://www.m1.com.sg/AboutM1/NewsReleases/2016/Results%20for%20nine%20months%20ended%2030%20September%202016.aspx.

Marston, Rebecca. 2017. "Smartphone Use Falls Among Young for First Time." *BBC News*, October 20. https://www.bbc.com/news/business-41805801.

Netflix. N.d. "Internet Connection Speed Recommendations." *Netflix Help Center*. https://help.netflix.com/en/node/306.

Realwire. 2017. "When Will Exponential Mobile Growth Stop?" Press release, October 10. https://www.realwire.com/releases/When-will-Exponential-Mobile-Growth-Stop.

Rosenthal, Marshal M. 2014. "4K Video Streaming Issues." *Videomaker*, February 16. https://www.videomaker.com/article/c12/17028-4k-video-streaming-issues.

Sherman, M., E. Peter, S. Sharma, M. Wilms, D. Locke, A. Dahlke, S. Stemberger, and M. Hitz. 2015. "Uncovering Real Mobile Data Usage and the Drivers of Customer Satisfaction." *Boston Consulting Group*, November 16. https://www.bcgperspectives.com/content/articles/telecommunications-center-for-customer-insight-uncovering-real-mobile-data-usage-drivers-customer-satisfaction/?chapter=4#chapter4.

Statt, Nick. 2016. "Pokémon Go Uses Little Data, But It's Still a Big Drag on Mobile Networks." *The Verge*, July 15. http://www.theverge.com/2016/7/15/12201418/pokemon-go-mobile-network-performance-data-use.

Sundaresan, S., N. Magharei, N. Feamster, and R. Teiseira. 2013. Measuring and Mitigating Web Performance Bottlenecks in Broadband Access Networks. The 2013 Internet Measurement Conference (IMC), Barcelona, Spain. http://conferences.sigcomm.org/imc/2013/papers/imc120-sundaresanA.pdf.

techUK. 2016. "Convergence and the Implications for Application Developers." Insight. http://www.techuk.org/insights/opinions/item/9505-convergence-and-the-implications-for-application-developers.

Tefficient. 2016. *USA, Latvia and Finland Combine: High Mobile Data Usage with Fast Growth*. Industry analysis #3 2016. June 29. http://media.tefficient.com/2016/06/tefficient-industry-analysis-3-2016-mobile-data-usage-and-pricing-FY-2015-final2.pdf.

Williamson, Brian. 2017. Mobile first, fibre as required: The case for "Fibre to 5G" (FT5G). United Kingdom: Communications Chambers, January. https://docplayer.net/36105634-Mobile-first-fibre-as-required.html.

Chapter 4
Technology Reaches Its Limits

Every generation has been both faster and more spectrum efficient than its predecessor. As shown in Chapter 2, data rates have grown approximately 10 times with each generation. Spectrum efficiency grew hugely, from 2G to 3G by 20 times, and then a further 2.5 times between 3G and 4G (so 4G is about 50 times more efficient than 2G). It might be assumed that 5G would deliver not only 10 times the growth but also significant enhancements in efficiency compared to 4G. Equally, the much lower growth level from 3G to 4G—when compared with 2G to 3G—suggests that it is getting harder to find ways to enhance efficiency. Broadly, growth in data speed is relatively easily achieved by increasing the bandwidth of the channel that devices use—for example, 4G uses channels of up to 20 MHz bandwidth, compared to 5 MHz channels for 3G (which, of course, requires more spectrum). But growth in efficiency is much harder to deliver.

Technical efficiency of data transmission is constrained by hard limits—the most relevant of which is the Shannon limit. This sets the maximum data transmission possible for a given channel and can never be exceeded. Hence, technical efficiency cannot keep growing. However, Shannon's limit applies directly to single users on a fixed pipe (e.g., copper or fiber) and only indirectly to wireless communications, where there are multiple users and complex communications channels. This means that understanding whether we are close to Shannon's limit is problematic, but most people who have studied it believe that there is little room for improvement. For example, Mogensen et al. (2007) state:

> We then use the adjusted Shannon capacity formula to predict LTE cell spectral efficiency (SE). Such LTE SE predictions are compared to LTE cell SE results generated by system level simulations. The results show an excellent match of less than 5–10% deviation.

This suggests LTE is within 10 percent of maximum theoretical capacity, and so further gains in areas such as coding and modulation are unlikely. This can be seen in the current struggle to find a "new radio" solution for 5G that is materially better than 4G.[1]

Assuming that there are no material gains to be had from basic channel efficiency, there remain two possible approaches to increased throughput—multiple antennas and small cells. Multiple antennas (often known as MIMO; specifically, multiple input, multiple output) provide two ways to increase cell throughput:

1. By creating multiple radio paths between the base station and the mobile, each of which can, in principle, carry data up to the Shannon limit.
2. By creating narrow beams and so reducing interference to other users, allowing each user to have a greater share of the cell capacity.

Smaller cells simply reduce the number of users per cell. If the cell capacity remains unchanged, it effectively allows more data per user. This chapter looks at these two possible approaches to enhancing efficiency. It then considers a radically different approach of keeping efficiency broadly constant but able to access huge new swathes of spectrum.

As well as the radio access, mobile networks consist of network cores that route traffic and manage connections. The 5G community has suggested that material improvements can be made by "virtualizing" this core—essentially enabling it to be implemented as a software load onto a general-purpose computing platform. The chapter concludes by asking, "is this revolutionary?" and, if so, what the resulting implications are.

MIMO

With MIMO, multiple antennas are deployed at both the base station and in the subscriber device.

There are two ways that MIMO can increase capacity. The first is "classic" MIMO, where multiple paths between the transmitter and receiver are created, with different data sent via each path. This requires as many antennas at the device as at the base station. The second is beam forming, where the antennas are

---

[1] For example, see Zhang et al. 2016, where a range of different competing radio technologies are evaluated. Results show that at low signal-to-noise radios (SNRs), where most cells operate, the existing 4G solution is superior to the new proposed solutions. At very high SNR values, which are rarely encountered, some of the new solutions have a small gain of about 10 percent. This suggests a risk that in real deployments, any new radio technology could actually be inferior to the current 4G solution.

used to form a more focused beam of radio energy that reduces the interference levels to others in the cell to enable greater capacity. Beam forming typically only occurs at the base station and does not require additional antennas at the terminal. So, for example, a 2×2 MIMO deployment would be an entirely classic MIMO, whereas a 4×2 MIMO deployment might use two of the base station elements for classic MIMO and two for beam forming.

If there are multiple radio paths from the base station to the device, then the differences between the signals received at each antenna can be used to extract additional information. In theory, the capacity gains from MIMO are equal to the lowest number of antennas on either the base station or the device. So (for example) a 2×2 MIMO, with two antennas at the base station and two at the device, is up to twice as efficient as non-MIMO transmission. A 4×4 MIMO is four times as efficient. However, 4×2 MIMO is only as efficient as 2×2 MIMO, unless the extra elements are used for other purposes such as beam steering. These gains come about because, in theory, each channel can be treated separately and can transmit data at up to the Shannon limit.

In practice, gains are much less than this, as radio conditions are less than optimal, and the necessary understanding of the channel conditions are imperfect. The worst case for MIMO would be a single direct radio path between the base station and mobile. In this case, the differences between the paths from the different antennas would be minimal, and hence they would interfere with each other if different data streams were encoded onto them. The best case for MIMO is when there are multiple reflections from nearby buildings and other obstacles, such that the path from one base station antenna to one of the mobile antennas is completely different to the path between another pair of antennas. The direct radio path case is more likely in smaller cells, while the case of multiple reflections is more likely in larger cells. Hence, MIMO gains are more likely in macrocells than microcells. As shown later in the chapter, the frequency bands associated with 5G will tend to result in smaller cells.

Because of imperfect radio paths the gains are not as large as theoretically possible. For example, moving from 2×2 MIMO to 4×4 MIMO has been shown in trials conducted by mobile operators to provide about 20 percent capacity gains Nokia 2017 and are not the 100 percent possible, theoretically speaking (capacity gains at cell edge are nearer 50 percent, but the overall average is 20 percent). With a law of diminishing returns, the gains from moving to 8×8 MIMO would be even less and, given the added complexity of placing eight antennas in the subscriber device, probably impractical.

Another practical problem is that for MIMO to be effective, antennas need to be well spaced so they can receive different signals. Typical recommendations for antenna spacing are for 5 to 10 wavelengths at the base station and for 0.5 to

1 wavelengths at the receiver for optimal performance. At 900 MHz, a wavelength is 30 cm, requiring 1.5 to 3 meter spacing at the base station. At 2.3 GHz, the wavelength falls to 12 cm, reducing spacing to 60 cm to 1 meter.

For a typical MNO, this means that employing a 4×2 or 4×4 MIMO will require additional antennas to be deployed on a base station site. At lower frequencies, physically separate antennas are needed to get sufficient spacing, whereas at higher frequencies, a single structure can be deployed with multiple antennas within it. Deploying additional antennas onto existing sites is generally very difficult owing to physical constraints and landlord permissions. Hence, increased MIMO is only practical in the higher bands, at 1800 MHz and above.

Given the difficulties in deploying more antennas at the base station and in the mobile and the diminishing returns from imperfect radio conditions, using classic MIMO in conventional frequency bands beyond the 4×4 already envisaged in 4G is unlikely.

The alternative is to use MIMO to form beams at the base station. These beams track subscribers, focusing the radio energy on them and so reducing the interference to others in the cell. A perfect beam-forming system could add hugely to the cell capacity, since each mobile would be able to access almost the entire radio capacity of the cell, independently from other users. However, to do this requires large and complex antenna arrays at the base station with expensive duplicated radio-frequency components. It is also very challenging to track the user as they move through the cell. At times, the user may pass behind obstacles, and thus the optimal beam might come from a reflection off a building rather than directly to the user. Finding these reflections, moving the beams, and keeping track of all subscribers in a highly dynamic environment is highly challenging and will result in the practical deployments having much lower performance than theory.

At present, beam steering is being considered for the millimeter wave bands (mmWave), discussed in more detail below. This is because antennas in these bands are much smaller, and so antenna arrays are more practical.

However, beam forming is a subject of much research. Bristol University and others have demonstrated early implementations of beam-steering systems designed to work below 6 GHz and with form factors that might be deployable. This is one area where there may be gains that can be realized by MNOs, perhaps even on 4G systems operating in bands such as 3.4 GHz. Much work remains to be done to determine whether such systems are viable in practical deployments. *If low-frequency beam forming that is implemented on macrocells proves commercially and economically viable, it might result in significant capacity enhancements.*

Small Cells (Microcells)

Historically, mobile radio systems have improved capacity through the addition of more cells. Each cell has the same capacity, so replacing one large cell by, say, 10 small cells, results in a tenfold capacity improvement. Cell sizes have reduced dramatically over the decades to the point where conventional "large" cells are often only about 500 to 1000 meters apart in city centers (implying a coverage radius of 250–500 meters). At this density, it is generally impractical to add more conventional cells, not only because finding sites for them can be highly problematic but also because they tend to increasingly interfere with each other, reducing the capacity gains that otherwise would be achieved.

An alternative is to use small cells. Small cells, or microcells, are those where the base station antennas are deployed below the level of the surrounding rooftops. As a result, their propagation is constrained by the buildings to the street canyons around them, giving them typical ranges of 100 to 200 meters. They can be either deployed as isolated cells in hotspots of high demand, such as shopping malls or stadiums, or as a more contiguous layer, providing coverage throughout a macrocell or across a wider area.

A classic approach of "densification" replaces one large macrocell with multiple smaller macrocells. This works effectively because the same frequencies that were in use in the initial macrocell can now be used in the new macrocells, enabling more reuse of the frequencies. But with small cells, the macrocell is never decommissioned, because the small cells rarely have complete coverage across the macrocell area and also because faster moving mobiles need to be supported on the macrocell. This means that the frequencies used in the macrocell are not directly available to the small cells, since the use of them in both would result in interference. As a result, the conventional wisdom that shrinking cell sizes improves capacity does not necessarily apply to small cells.

Within 4G, there are three approaches that can be adopted to resolve this issue:
1. Take some of the frequencies from the macrocell and deploy them in the small cells. This removes interference but results in a reduction in capacity on the macrocell for which the small cells have to compensate. If there are few small cells, they may be unable to compensate for the loss of macrocell capacity, and the overall capacity across the area of the macrocell may fall.
2. Attempt to use the same frequencies in both and try to manage interference by handing users over to either the small cell or macrocell depending on signal strength. However, this can result in a poor experience for customers toward the edge of the small cell where the signal strength is near equal from both; hence interference is problematic regardless of the cell selected.

3. Prevent interference between a macrocell and microcell sharing the same spectrum. MNOs are able to deploy interference cancellation techniques, such as enhanced intercell interference cancellation (eICIC), where macrocell base stations stop transmission on particular resource blocks (i.e., combinations of timeslots and frequency slots) which are used by the small cells. Essentially this is a way to take part of a frequency from the macrocell and vary how large this part is according to dynamic need.

A problem with isolated small cells targeted at hot spots is the importance of the base station location. For example, if a small cell is aimed at carrying traffic from a shopping mall, it will need to cover the majority of the mall. Deployment at one end of the mall might result in most of the wireless traffic being carried on the macrocell and the small cell being underutilized. However, finding ideal locations is not always possible.

Finally, in-building penetration of small cells is variable. If the small cell is nearby and can "see" well into the building, then the penetration is good. However, as it moves further away, the angle of the beam becomes more oblique to the building and the penetration falls rapidly. In-building coverage tends to be provided by the macrocell, which is more likely to have a better angle of view into the building. Also, with their low height, small cells can only illuminate the lower floors of buildings—typically the ground and first floor. In dense cities, most buildings have many more floors than this, so much of the in-building traffic cannot be served by outdoor small cells.

Detailed modeling of a variable number of small cells (Webb n.d.) shows that the first small cell deployed within a macrocellular sector reduces the capacity of the combined macrocell and small cell. This is because it takes some of its frequency allocation from the macrocell, reducing its capacity by more than the capacity it is adding. The second and third small cells can add significant capacity as they can reuse the same frequencies being used by the first small cell and so do not materially increase interference. This is particularly true if they are deployed in hot spots and not overlapping with each other. Such a hotspot deployment of three small cells can increase the sector capacity by around 50 percent, as shown in Figure 4.1.

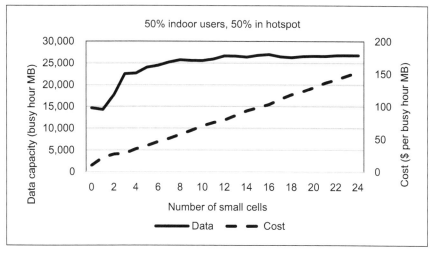

Figure 4.1: Capacity and Cost of Small Cells

Going from four to about 10 small cells provides smaller incremental gains. Gains are limited because the small cells cannot serve many of the indoor users and so do not attract large volumes of traffic. With 10 small cells in a sector, the overall capacity increase is typically less than 80 percent depending on the distribution of traffic between indoor and outdoor users.

Beyond this, capacity is essentially static, as additional small cells increasingly overlap with existing small cells, and the interference created is equal to the capacity added.

The combined costs for the macrocell and the microcell, measured in terms of the units of data carried, rise throughout, as they are three times higher for three small cells, six times higher for nine small cells, and over ten times higher beyond this amount. Hence, small cells are an expensive way of providing additional capacity.

These results suggest that deployment of a few small cells in traffic hotspots can be effective, but that there is little point in deployment beyond the number of hotspots in a sector—and more generally beyond about three or four small cells per sector. In any case, the maximum gains in capacity from small cells, even with a complete layer, are less than 100 percent capacity increase. This comes at a disproportionate cost that may be hard to justify in the absence of ARPU increases (Webb n.d. provides considerably more detail on this issue).

Hence, *using small cells outdoor provides less than 2 times capacity improvement and comes at a high cost.* This is a conclusion that is discussed in Chapter 5, when we consider the economics of the mobile network operator.

The situation changes dramatically if the small cells are placed within buildings. In this case, they can effectively target the traffic and the interference they generate in the macrocell is limited by the effects of the outer wall of the building. By using indoor small cells, huge increases in capacity are possible.

MNOs have tried to deploy indoor small cells for decades. Various initiatives such as picocells and femtocells have been tried with limited success. MNOs find it logistically challenging to deploy in-building, as they need the permission of the building owner and access to most floors in nearly every building. Owners are typically not interested in hosting a single MNO, and providing space and access for multiple MNOs is rarely worth the time and effort. Instead, building owners tend to deploy their own Wi-Fi solutions, and mobile users are transferred to these solutions for all data usage, only remaining on the cellular system for voice. With this approach becoming widespread, the incentive for MNOs to deploy within buildings becomes weaker, as their cells would typically only carry voice traffic. The implications of this are considered further in Chapter 8.

There are regional differences in small cell deployments. In some Asian-Pacific countries there are many more small cells per person than in Europe and the United States. This appears to be predominant because site costs are lower and backhaul is more readily available in places such as China and Japan. It might also be because of the size and density of cities like Tokyo and Beijing. This would suggest a somewhat higher network capacity per user in these countries than is average.

mmWave

The wavelength of radio waves is related to their frequency by the formula $v = f\lambda$, where v is the speed of light, f the frequency, and λ the wavelength. Above 30 GHz the wavelength falls below 1 cm, and hence the bands are sometimes termed millimeter waves (mmWaves). This terminology is shorthand for the idea that in 5G these higher frequencies bands could be deployed for the first time. The bands considered for 5G start at around 26 GHz, which are strictly not mmWaves, but there is clear division between these bands and those currently used for mobile communications, which are all below 4 GHz.

The rationale for moving to mmWave bands is because there is potentially much more spectrum available. To acquire 1 GHz of spectrum below 4 GHz would mean taking 25 percent of the band. Acquiring 1 GHz in the mmWave bands between 25 GHz and 100 GHz is only 1–2 percent of the available bandwidth. The large amounts of spectrum can result in much improved cell capacity without the need for improvements in spectrum efficiency and can also deliver blindingly fast data rates of well over 1 Gbps.

However, there is a good reason why these bands have not been used for mobile communications to date. Propagation losses are extremely high and as a result the range is limited to typically around 100 meters for mobile use (FWA may be viable up to 1 km, as explored further in Chapter 8). This implies that mmWave solutions are always based on small cells and, as discussed earlier, have the disadvantages of poor access to indoor users as well as the high costs associated with finding sites and backhaul.

The poor range can be somewhat improved with directional antennas that focus transmissions on specific users. However, the narrow antenna patterns will result in dynamic variations in coverage, signal quality, and channel quality with slight movements in the user's device or with objects in the environment, such as passing vehicles. At the same time, signal blockage from obstacles can greatly reduce the beam coverage. This may lead to frequent handover requirements between different beams in order to provide sufficient coverage and connectivity (Safjan 2016).

Even getting connected in the first place is challenging, since the base station and the user can only connect either when beams are correctly aligned or when the user is so close to the base station they can be heard without needing beam forming (Giordani et al. 2016). Correctly aligning the beams when the user is unaware of the base station location and the base station is unaware of the presence of the user[2] is challenging; it typically requires "search" techniques, where beams are periodically scanned around the cell. This adds delay to finding the user, often in the region of one to two seconds, which does not fit well with providing a low-latency solution. If a user is in a car traveling at, say, 20 miles per hour, with cells of a radius of 100 meters, then a reconnection could be needed every 20 seconds or so. Alternatively, the network can wait until the user gets within 30 to 40 meters of the base station, where communications without beam forming might be possible, but this restricts the effective range of the cell, reducing its coverage area by 75 percent or so.

These technical challenges may give rise to increased need for control plane traffic and system complexity and may even make reliable use of the bands impossible in some environments.

---

[2] It is possible to use the low-frequency cells to provide information to the user as to the mmWave cell location and for GPS on the user's phone to tell the network the location of the user, but often the optimal beam is not a direct line between the user and the base station, so this approach fails. It also requires significant battery consumption to maintain GPS tracking and multiple conversations with low frequency and high frequency cells.

Limitations in propagation capabilities at mmWave have prompted some to consider that such bands may be best suited to coverage of small areas with high capacity requirements (e.g., stadiums, music concert halls).

Availability of sites is a major issue for mmWave deployments. A contiguous deployment throughout a city would require sites every 200 meters or so. Each site would need to be able to support potentially large antenna structures. Suitable sites might be lampposts, but even these will typically only support a single operator, suggesting that a single network might need to be deployed and shared across all MNOs. Lampposts typically do not have fiber backhaul available, making it expensive to connect the cell back to the network. Many have noted that a change in the building regulations may be needed in most countries to make small site acquisition simpler, but there have been such calls for decades with little, if any, action. With these issues unsolved, deployment will be expensive and will likely be constrained to outdoor dense city areas. This means that they cannot be used to deliver new services or applications, since subscribers would want these to be ubiquitously available. Instead, they can only deliver greater capacity. But most capacity is needed indoors and can be readily delivered with Wi-Fi solutions.

As noted previously, there are greater numbers of small cells in some Asian-Pacific countries. This might lower the deployment costs of mmWave solutions in these countries. Equally, it might make mmWave solutions less necessary since current capacity levels will be high.

There are further issues. As discussed earlier, beam forming is very challenging; much research is likely to be needed before it works well in practice. Consumer devices have never been made in these frequency bands before, so there will need to be widespread work on cost-reductions for key components. Practical issues abound.

There is also no agreement on the frequency band to use, with the United States proposing 28 GHz, Europe 24 GHz, and other countries 33 GHz. Without global harmonization it will be harder to produce low-cost devices and more likely that manufacturers and operators will pause, awaiting more certainty.

At a recent debate (Cambridge Wireless 2016a,b) there was general consensus from manufacturers, operators, and many others in the industry that the mmWave solution might not appear until 2025 or later. However, as shown in Chapter 3, by this stage growth in demand may have leveled off, with little need for an expensive and complex solution of this sort.

In summary, we are still some way from the point where it will be clear what (if any) role mmWave systems will play. Given the high interest of mmWave in countries such as China and Korea, it may be that the first indication of the difficulties and uses of mmWave solutions will appear in those markets.

Full Duplex

A relatively new proposal that might find use within 5G is the concept of "full duplex." Conventional communications make use of different radio paths for the downlink (from the base station to the terminal) and the uplink (from the terminal to the base station). For example, in 4G these typically occur on separate frequency bands in an approach known as frequency division duplex (FDD). If these could occur on the same band and at the same time, then there is a potential for a doubling of capacity.

However, there are myriad problems with this. The first is that the balance of traffic across the downlink and uplink is very uneven, with users typically downloading about 10 times more than they upload. This has been reflected in recent 4G systems where the bands are balanced accordingly. In such a system, enabling the uplink to be sent in the same band as the downlink only provides a 10 percent capacity gain.

The second problem is that when these transmissions are simultaneous, interference occurs. Operators will suffer from interbase station interference as well as challenging design issues within the handset that will add to the cost. While this is not the place for a further exploration of the issue, this essentially means that a base station will now be listening to a mobile phone while another nearby base station will be transmitting in the same band. Because base stations transmit at higher powers than mobile phones and are in elevated locations, the interfering signal from the other base station could be much stronger than that from the mobile, rendering reception impossible. Overcoming this requires reducing transmission rates or powers, negating many of the gains.

In summary, while full duplex has proven to be a fertile research topic, it is hard to see how it can add significant capacity in most cellular networks.

Core Network Evolution

Mobile networks consist of two key parts—the radio access and the core. This chapter has focused up to now on radio access, which typically comprises 90 percent of the network cost with its thousands of base stations and radio elements. The core connects these base stations to switching and routing points and provides functions such as billing and location management. Historically, the core has been implemented on bespoke hardware platforms such as mobile switching centers (MSCs) and packet gateway routers. However, as general-purpose computing platforms have become more widespread, there is a move toward "virtualization" of the network by implementing it as a software load on cloud-based computing platforms. In this case, there would be no "actual" core network, just a function that delivered the same outcome running on third-party hardware. This approach is termed network function virtualization (NFV).

NFV is already being implemented in 4G systems. Its primary benefit is cost saving for the MNO both through reducing the hardware cost of the core network and through reducing the cost of implementing new functions. However, with the core network typically comprising 10 percent of total network cost, even if NFV halved the core network cost, the overall savings to the MNO would not be transformational. There is also a risk that implementing new software in the core could result in instability or unexpected failure, potentially leading to complete network outages. MNOs will want to proceed with great caution.

NFV has also become associated with the concept of network slicing—the ability to configure network architectures dynamically according to varied use cases (e.g., low-latency usage, high capacity cases, low bandwidth machine-to-machine communications). In this concept, a single network can carry traffic of multiple different types, setting aside the necessary resources for each and so ensuring that each has the appropriate quality of service. However, the need for network slicing is far from clear. Network problems arise from congestion—when the demand for resources exceeds the supply. Congestion in mobile networks is almost invariably in the radio link; hence the desire to improve spectrum efficiency. The backhaul from cells back to the core network can be dimensioned to be larger than the cell capacity so it never gets congested, and core networks are typically specified at traffic levels above total network capacity. In the core, capacity is relatively low cost, as fiber optic links are used with near-infinite capability and processing resources can be rapidly scaled as needed. Hence, there is little need to "slice" the network into separate elements. Instead, the radio interface needs to be flexible enough to prioritize certain types of traffic as congestion increases, and some devices need to be able to defer their transmissions until network usage drops. This is a radio function—not a network function. There seems little benefit in network slicing.

Overall, NFV is a useful development, already enabled within 4G, which can provide cost savings of the order of a few percentage points to MNOs and might enable more rapid introduction of new services.

Heterogeneous Networks

The current communication environment is made up of a number of different networks. Key among these are the cellular networks and Wi-Fi provision. In addition, emergency services often have their own network; IoT networks may proliferate in coming years. Increasingly cellular and Wi-Fi have become integrated within the handset, with most cell phones switching seamlessly between them. However, there has been little integration at the network level, with the result that voice calls cannot be handed over between cellular and Wi-Fi and calls to a cellular number may not be able to be redirected to a voice-over-Wi-Fi service.

One of the concepts of 5G is that there could be tighter integration between these different networks. This is termed heterogeneous networking, or hetnet for short. Devices that are able to move seamlessly to the best network should be both more efficient and able to offer a better service to the user. Various concepts have been suggested to achieve this, including separating the data streams used to control the device and those used to send data, known as control/data separation. Such separation would allow the device to be controlled from one network (typically the cellular network) while it accessed data from a different network (e.g., Wi-Fi). It also makes handover easier because there is a constant channel to the device, although it does require the device to be dual connected. But this is still a work in progress.

The hetnet concept appears sensible in principle, but the benefit is less clear.

First, it would only appear to apply to cellular and Wi-Fi. Emergency services typically have multiple devices to access different networks and are looking to move onto 4G cellular networks, as discussed further in Chapter 7. IoT devices would be very unlikely to move across to a different network, as they are simple devices optimized to run on single low-power networks.

Next, from an end-user's point of view, this integration is already in place. Devices do move from one to the other, and handsets can already have a voice call on cellular while simultaneously downloading data from Wi-Fi. As mentioned above, there are a few cases that could be improved, but the major benefit has already been captured.

Finally, this would not appear to be in the interests of the MNOs. It would result in less traffic on their network as more traffic is off-loaded onto Wi-Fi and might make concepts like Google's Project Fi[3] simpler to introduce. While 5G does not need to be a solution solely for the MNOs, it is they who will have to introduce

---

[3] Project Fi allows users to sign up to Google instead of an MNO. Then, their phone attempts to use Wi-Fi for all activities, falling back to an MNO partner if that fails. From the user's viewpoint, they have a lower-cost monthly subscription with better coverage since Google can have multiple MNO partners, effectively enabling a form of national roaming. Google can deliver the service for less cost as most data flows across free Wi-Fi connections. The MNOs lose out as they now only get wholesale roaming fees rather than subscriptions.

the core network changes to enable hetnet operation. It is hard to see why they would bear the cost of an upgrade that might reduce their revenue.

A Route to Structural Change

While NFV and hetnets may have limited benefit to the MNOs, there may be unintended consequences leading to a restructuring of the industry. At present the MNOs run radio access networks (RANs) comprised of the masts (often leased from mast providers such as American Tower and Crown Castle) and base stations and core networks comprised of the switching and control. Over-the-top (OTT) providers such as Google and Facebook then deliver services. If NFV moves the core network element into the cloud where the OTT providers already often reside, it will open the door for a restructuring where MNOs own the RANs and OTT providers own the core and services.

This might be a more natural fit, since the OTT service could then be coupled more closely to the network capabilities, and new OTT service features that would benefit from changes to network protocols could be quickly introduced. Whether there would be a single "core and OTT" company per RAN, or multiple companies each with its own OTT service, remains to be seen.

In such a world, the role of the MNO is much diminished. They become the wholesale provider of RAN facilities, which is an even smaller role than when they provide wholesale capacity to mobile virtual network operators (MNVOs) today. Such positioning makes it easier to consider consolidation among MNOs since there is now vibrant competition provided above the RAN by the multiple core/OTT operators. Network slicing might facilitate this. By enabling the RAN to be sliced among the different OTT operators, and by enabling any core elements to be sliced among multiple OTT providers, a more widely shared network could be created.

Hence, in this future, there might be only one or two RAN providers after a process of consolidation among the MNOs, along with a vibrant provision of communications services from companies such as Google or Facebook, who offer a fully featured communications portfolio built on these RANs and also other communications networks such as Wi-Fi access points. At this point, the hetnet capabilities become helpful, allowing such providers to make the best use of cellular, Wi-Fi, and any other communications resources. Change like this is already underway, in particular in the United States, where the industry structure continues to evolve. The fifth generation might be more about structural change in the industry than new features for users, which will be discussed later.

NFV is also likely to have an impact on the supply industry. MNOs no longer need to source their core equipment from traditional vendors such as Ericsson and Nokia. Instead they can buy equipment from companies like Cisco or make use of cloud computing (where cloud providers tend to buy from Cisco or others). They will no longer need to subscribe to software updates from the vendors either, and this will open the door to smaller players or other new entrants that will provide competitive supply. As such, the traditional vendors will not relinquish their position readily and may block the deployment of NFV by delaying the changes needed to existing core networks to enable it.

Conclusions

This chapter shows how technology has improved dramatically over previous generations but has now reached a point where further improvements are hard-won. This broadly means that advances can become more expensive in the form of many more antennas at the base station and in the device, many more small cells, or dense deployments in completely new frequency bands. All of these are uncertain; some are untried, and some will require substantial further development. The advent of 4G effectively provided about 2.5 times the previous capacity at very little extra cost. The same will probably not happen for 5G. Capacity enhancements appear to be below two times (industry estimates suggest perhaps only 1.2 times) and come at a very high cost, although there is some possibility that after much research, beam-forming antennas in bands below 4 GHz might lead to gains in capacity. The implications of capacity gains being hard to find will be considered in the next chapter, which looks at network economics.

Chapter 4 References

Nokia 2017.https://onestore.nokia.com/asset/200187
Cambridge Wireless. 2016a. "Debate 1: Technology Readiness." The CW 5G Debate in association with the NIC held at the Shard, London, United Kingdom, October 24. https://www.cambridgewireless.co.uk/media/uploads/resources/Debates/24.10.16/5G_Debate1_24.10.16_FinalTranscript.pdf.
Cambridge Wireless. 2016b. "Debate 2: Business Cases." The CW 5G Debate in association with the NIC held at BT Tower, London, United Kingdom, November 8. https://www.cambridgewireless.co.uk/media/uploads/resources/Debates/08.11.16/5G_Debate2_08.11.16_FinalTranscript.pdf.
Giordani, Marco, Marco Mezzavilla, and Michele Zorzi. 2016. "Initial Access in 5G mmWave Cellular Networks." *IEEE Communications Magazine*, November, 40–47.
Mogensen, P., Wei Na, I. Z. Kovacs, F. Frederiksen, A. Pokhariyal, K. I. Pedersen, T. Kolding, K. Hugl, and M. Kuusela. 2007. "LTE Capacity Compared to the Shannon Bound." Presented

at 2007 IEEE Sixty-Fifth Vehicular Technology Conference (VTC2007), Dublin, Ireland, April 22–25. https://ieeexplore.ieee.org/document/4212688.

Safjan, Krystian, ed. 2016. "Architectural Aspects of mm-Wave Radio Access Integration with 5G Ecosystem." mmMagic White paper, April 14. https://bscw.5g-mmmagic.eu/pub/bscw.cgi/d100702/mm-wave_architecture_white_paper.pdf.

Webb, William. N.d. "Limits of Small Cells in Dense Networks." Paper, Webb Search Consulting. http://www.webbsearch.co.uk/wp-content/uploads/2013/09/Small-Cell-Paper-for-Journal.pdf.

Zhang, Xi, Lei Chen, Jing Qiu, and Javad Abdoli. 2016. "On the Waveform for 5G." *IEEE Communications Magazine*, November, 74–80.

Chapter 5
Economics Reaches Limits

The move through the mobile generations has resulted in enormous benefits for subscribers. For virtually unchanged monthly subscriptions (often known as average revenue per user—ARPU), subscribers have gone from voice and SMS packages to the ability to send over a gigabyte (GB) of data at rates of over 10 Mbps. Subscribers have become used to getting ever larger bundles of data for the same subscription cost.

The last few decades have been one where consumers have benefited immensely. Governments have come to expect their resident MNOs to deploy the latest network generation without any explicit support and indeed with large payments to the government via auction fees for access to the spectrum.

MNOs, and the associated ecosystems of suppliers and others, are substantially listed companies. They have duties to make good returns for their shareholders and suffer a value loss if it is perceived that they are less profitable than other major companies. This chapter discusses whether the bonanza enjoyed by subscribers can continue into the 5G era or whether profitability concerns will finally have an impact.

MNO Performance

In the era of 2G, MNOs were some of the most profitable companies on the stock exchange. Revenues were growing strongly on the back of surging subscriber numbers, and costs were reasonably constrained with few having to pay for their spectrum. Competition was often muted with typically only two operators in the marketplace. While MNOs had to acquire thousands of sites, there were plenty of potential locations, and landlords had yet to realize that they could charge relatively high rental values.

Profitability dropped in the era of 3G. Many operators incurred substantial fees at spectrum auctions beginning about the year 2000, which is about the same time that more competition entered into the market (with many countries increasing the number of native MNOs to four or more). Subscriber growth slowed in developed countries as most users acquired mobile phones. MNOs reacted by cutting back on activities such as R&D and through regional consolidation, with companies such as Vodafone and Telefonica building portfolios around the world.

Profitability fell further with the advent of 4G. MNOs were required to pay for more spectrum—albeit typically only at about 10 percent of the level of the 3G auctions—and needed to invest in new network equipment. Some tried to raise subscription levels in response, charging more for 4G services, but this had very limited attraction to consumers and most quickly "gave away" 4G at no additional cost to the subscriber. MNOs now accept that subscribers will pay little more for new services and will quickly migrate to the network offering the lowest cost.[1] Figure 5.1, produced by the GSMA (the MNOs' industry body), shows the trend and predictions for revenue growth.

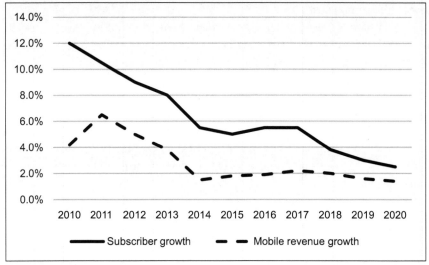

Source: GSMA 2016.
Figure 5.1: Predictions of Growth and Revenue

As can be seen, revenue growth dipped below 2 percent in 2014 and looks set to remain at that level for the foreseeable future. This figure is consolidated across the world. In developed countries growth is often negative but balanced by stronger growth in countries such as India and China. Few in the industry currently expect to be able to charge subscribers more for 5G services. Revenue is expected to grow more slowly than GDP from 2017 onward.

In 2016 the United States saw large declines in ARPUs of 7 percent (FCC 2017). This is shown for the last 20 years in Figure 5.2 and then for the last five years in Figure 5.3.

---

[1] A few MNOs managed to increase their ARPUs through careful management of their subscriber base, but this can just shift the low-ARPU customers to other MNOs.

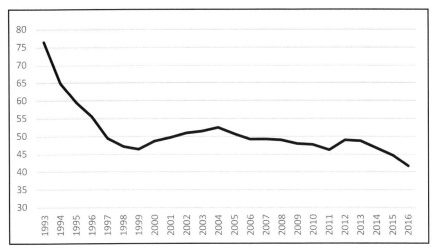

Source: FCC 2017.
Figure 5.2: US Average Revenues per User, 1993–2016

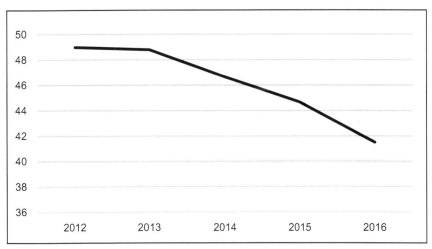

Source: FCC 2017.
Figure 5.3: US Average Revenues per User over the Last Five Years

This can be attributed to competition finally having an impact on ARPUs that have been unusually high by global standards. It is also partly due to unlimited data bundles that prevent further upselling and encourage businesses to trade down their connection to a consumer-price connectivity package.

So this is an industry with little growth; in developed countries, revenue is in decline.

Figure 5.4 also from the GSMA shows what has happened to earnings before interest, tax, and depreciation (EBITDA).

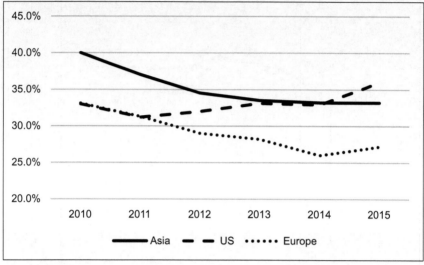

Source: GSMA 2016.
Figure 5.4: EBITDA Figures by Region

This figure shows a mixed picture, with European operators suffering while those in the United States returned to EBITDA growth. But to put this in perspective, the enterprise value divided by EBITDA (a common metric of comparison) for US wireless stocks[2] was 7.6 versus an average across all industries of 14.7 in January 2016.[3] The ratio at a global level was very similar.

Figure 5.5 shows that investment is far from over. While there was an investment peak in 2015 relating to the 4G rollout, there is not much decline to 2020—and these figures do not assume any 5G investment, which the GSMA does not expect to take place before 2020.

---

[2] This is likely to include a wider community than just the MNOs.
[3] See Damodaran 2017; the enterprise value (EV)/EBITDA column where "telecom (wireless)" has a value of 7.6 against 14.4 across all sectors.

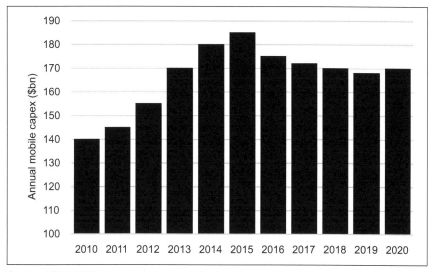

Source: GSMA 2016.
Figure 5.5: Mobile Sector Capex

However, this capex is broadly not predicated on additional cell sites but rather on "business as usual" network upgrades. For example, the number of cell sites in the United States has been virtually static for the last four years, as shown in Table 5.1.

Table 5.1: US Cell Site Count

	2013	2014	2015	2016
AT&T	61,800	71,768	66,500	67,000
Sprint	55,000	55,000	55,000	50,000
T-Mobile	63,879	61,079	57,971	59,417
Verizon	46,655	50,065	54,000	58,300
US Cellular	6,975	6,220	6,297	6,415
Total	234,309	244,132	239,768	241,132

Source: FCC 2017.

If 5G required significant expenditure, such as more cells, then the EBITDA ratios would trend downward well below the average across all sectors.

Over the last five years the value of the large MNOs has increased around 30 percent. During the same period, the Standard and Poor's (S&P) 500 grew 83 percent, although the Financial Times Stock Exchange (FTSE) 100 only grew by around 20 percent. Companies like Apple, Google, and Facebook grew by 400 percent. As such, MNOs do not appear to be an attractive place to invest.

MNOs have sought to change these dynamics through mergers to enable cost savings and reduce competition, which might allow increased prices. However, in almost all cases these have been blocked by regulators who fear a reduction in competition. These actions have locked the industry into a zero-growth zero-sum game, where staying alive requires doing enough to prevent subscriber churn (attrition) while spending as little money as possible. *For MNOs, this is a moribund industry in increasing need of change.*

5G Costs

Figure 5.5 shows that investment in 4G had been substantial for the industry. But 4G only required an upgrade to existing base stations—with new hardware and in some cases, new antennas. Most MNOs did not deploy significant numbers of new cells,[4] nor did they move to small cells or other similar architectures. As far as moving from one generation to another was concerned, this was about the lowest possible cost transition.

The costs of 5G are far from clear, and, as will be shown in Chapter 7, there is still much uncertainty about the form they will take. However, earlier chapters have shown that:
1. 5G will require new spectrum that will have auction fees.
2. 5G may require deployment of new antennas at base station sites as well as equipment upgrades, and this will be as expensive as the 4G transition, if not more.
3. If large-scale small-cell deployments are involved (e.g., as part of the mmWave vision), then costs could be very significant.
4. Handset costs may also rise as a result of increased complexity, and MNOs may end up bearing part of these costs if they subsidize handsets.[5]

---

[4] For example, in the United States, prior to 4G there were some 253,000 sites (CTIA 2017). This grew 20 percent by 2013 but has stayed approximately flat since then at around 305,000.
[5] While the trend has been away from handset subsidy, if MNOs wish to get a new capability (such as an mmWave operation) into the subscriber base, then they may need a subsidy to speed the handset introduction cycle.

Expectations of ARPU Growth

Figure 5.1 shows that mobile revenues are expected to only grow at about 2 percent a year—below the rate of subscriber growth. This implies that ARPUs are falling. Others agree. For example, Ovum states:

> In nearly all markets, ARPU will fall over the forecast period as a result of increased competition and a rise in take-up of non-phone connections, which will constitute a growing proportion of net additions over the forecast period. This is particularly the case in developed markets, where smartphone penetration is already high and where most operators' net additions will be driven by customers moving from one operator to another, rather than from market growth.
>
> The slowdown in new-customer acquisition will drive pricing competition among operators, which will in turn push down ARPU. A fall in flat-rate monthly tariffs and the introduction of shared data plans will encourage prepaid-to-postpaid migration and churn within the existing customer base. Tablets and other connected devices along with VAS services [value-added services, or noncore services] that aim to capitalize on these new types of connections will increase the mix of non-phone connections in the short term. Because these attachment subscriptions tend to carry lower ARPU than phone subscriptions, they will drive down overall ARPU. They will also probably boost churn somewhat among existing customers. Increasing availability of Wi-Fi might encourage some customers to use their cellular-capable portable devices exclusively over Wi-Fi. And a general lack of consequences (such as early-termination fees) for downgrading service might also encourage such behavior. (Ovum 2015)

Ovum picks up on another constraint for the MNOs. Any attempt to increase prices by one MNO risks driving subscribers to their competitors. Any attempt by all MNOs to increase prices will drive subscribers onto Wi-Fi—a good substitute for most subscribers, most of the time, especially with voice apps such as WhatsApp.

It is worth recalling why this is happening. MNOs are unable to raise prices because consumers broadly do not value any increases in the data rates and data volumes that are currently offered. While increased data volumes are nice to have, they are generally not worth the cost. This is precisely what the prediction would be from the analysis in Chapter 3, where we noted that current speeds are sufficient. Data requirements are growing, but these can also be met through Wi-Fi connectivity.

The industry is locked into a game akin to the prisoner's dilemma. MNOs' best option is for none of them to invest in new infrastructure, keeping the competitive position stable and minimizing their expenditure. However, if one breaks rank and, as a result, can offer a better service, then that MNO might attract subscribers from the others. The loss of revenue from these subscribers could be significant. Because there is a relatively long lead time on network investment, all

MNOs hedge their bets by acquiring spectrum and putting mechanisms in place that would allow them to react quickly to others. But if MNOs conclude that there is little value in 5G that would result in subscribers moving to the first MNO who deploys it, they will all be more inclined to sit back and not invest.

Table 5.2 shows the outcome where consumers value 5G highly, and some switch from operators that choose not to offer it.

Table 5.2: Investment Choices when Consumers Value 5G Highly

		Operator A	
		No 5G investment	5G investment
Operator B	No 5G investment	No revenue decline	A loses 10% in costs, B loses 20% of revenue
	5G investment	B loses 10% in costs, A loses 20% of revenue	A and B lose 10% of costs

Here it is assumed for illustration that 5G investment equates to 10 percent of revenue over the deployment period, and 20 percent of subscribers churn to the operator offering 5G services. As can be seen, the best option is for neither Operator A nor B to invest. But the worst option for A is not to invest when B does and vice versa. Prisoner's dilemma game theory suggests they both end up investing.

But now consider the situation where there is little in 5G that is attractive to consumers and so perhaps only 5 percent switch to a 5G operator as shown in Table 5.3.

Table 5.3: Investment Choices when Consumers Value 5G Highly

		Operator A	
		No 5G investment	5G investment
Operator B	No 5G investment	No revenue decline	A loses 10% in costs, B loses 5% of revenue
	5G investment	B loses 10% in costs, A loses 5% of revenue	A and B lose 10% of costs

Here, operators are clearly better off not making the investment. Game theory shows that neither invest.

What might the percentages actually be? So little is known about 5G that it is hard to put any cost estimates on it, but back-of-the-envelope estimates are possible. For a service to be compelling enough that it would cause significant subscriber migration, the service would have to be offered ubiquitously. This might involve, say, a rollout of a new radio solution at 3.5 GHz and the deployment of mmWave systems in very dense urban areas.

The 3.5 GHz system would be deployed on all existing base stations. Taking the United Kingdom as an example—say an MNO has around 20,000 base stations and the upgrade cost per base station—including equipment, site visit, new antennas, and commissioning—might be around $50,000 per site. In addition, since the range of 3.5 GHz is low, fill-in sites might be needed in some critical areas, perhaps amounting to a 25 percent site increase, or 5,000 sites. These might cost $150,000 per site, including site acquisition, mast building, amortized operational expenditure, etc. There would also be spectrum acquisition costs at auction of perhaps $600 million. The total for the 3.5 GHz deployment is then $2.35 billion.

The mmWave deployment might need around 5,000 small cells to cover dense areas of cities such as London and Manchester (indeed, some estimates put the numbers as high as 42,000 just for the center of London—see UK National Infrastructure Commission 2016). At around $50,000 per site, including equipment, installation, and amortized rental, and at a spectrum cost of $100 million, the total comes to $350 million. Core network upgrades might be a further $150 million.

The total of all these costs is $2.85 billion. This equates to a cost per subscriber of $190, which aligns reasonably well with a European Commission study.[6] This study suggested a cost of €145 per subscriber—around $155—but did note that 5G costs might be higher than this due to the scale of the ambition and that operational costs may be significant each year (these costs are not included here). Further, if the higher numbers for suggested small cells (UK National Infrastructure Commission 2016) are used, then these costs rise substantially.

A typical UK operator has around 15 million subscribers. The average revenue per subscriber per month, after handset subsidy, is around $20. Of this, $10 is direct subscriber costs. The remaining $10 makes a contribution to the fixed costs and can generate profit (although most MNOs are unprofitable or only marginally profitable). Hence, the lost revenue amounts to around $120 per subscriber per year. Assuming that the 5G investment is for a 10-year period, then the cost of the investment is equivalent to the revenue contribution from nearly 15 percent of the subscriber base.

This implies that an MNO would prefer not to make the investment in 5G if they expected less than 15 percent of their subscribers to churn to other operators

---

[6] See Figure 5 in Section 6.2 of Tech4i2 et al. 2016.

who had deployed 5G. Alternatively, if they were concerned that more than 15 percent would churn, then they would make the investment as a defensive play to avoid an even greater loss. Note in both cases, the MNO loses money.

These numbers are highly approximate, but they allow some estimation of 5G's attractiveness needed to persuade MNOs to deploy it. Might 5G be so attractive that 15 percent or more of subscribers permanently leave an MNO that does not offer it? Or might subscribers prefer an MNO that can offer subscriptions that are around 8 percent lower as a result of not having made these investments? Previous chapters have demonstrated that faster speed or more capacity is very unlikely to be a compelling reason for a subscriber to prefer 5G. Chapter 7 shows that there are few, if any, new services that cannot be provided with 4G. On this basis, a rational MNO would not invest in 5G and instead seek to compete on price.

Without a new "killer application," there is no expectation of ARPU increases after 5G deployment and little incentive for MNOs to deploy it other than as a means to expand capacity in congested areas.

New Verticals

By 2017 the industry was focusing on "verticals" as a way to generate growth as growth in consumer ARPUs seemed unlikely. The essential thinking was that particular industries such as automotive or manufacturing might be a new source of revenue due to the new services 5G might provide. In particular, it was thought that the ability to "slice" the network (as discussed in Chapter 4) might enable higher quality of service delivery, which would be attractive to these verticals.

However, most were very vague on what the verticals were. Automotive was frequently mentioned, along with healthcare and manufacturing. It was also very unclear as to where the additional revenue might come from.

Each of these possible verticals is discussed further in Chapter 7. It is sufficient to say here that the likelihood of significant new revenues appears weak. This explains why proponents of 5G verticals have, to date, been unable to demonstrate the robust business cases that would be needed to convince the MNOs to deploy 5G networks.

Conclusions

This chapter has shown that the economics of the mobile industry have changed substantially over the decades. From the time of 2G, when the MNOs were some of the most profitable listed companies, these MNOs have fallen to the point where they are underperforming in "all sector" benchmarks by some 50 percent. Revenues are not expected to rise, while investment is anticipated to continue at relatively high levels.

Absent growing APRUs, the only rationale for MNOs to invest in new technology is to prevent subscriber churn to their competitors. This threat has resulted in them moving quickly to deploy 4G, which has material benefits for subscribers. But without any clear benefits from 5G there is limited incentive for MNOs to upgrade their network.

The following chapters will discuss why most in the industry appear to believe that 5G will be transformational and more critically examine what 5G might actual entail.

Chapter 5 References

CTIA. 2017. *The State of Wireless 2017*. Washington, DC: CTIA.

Damodaran, Aswath. 2017. "Enterprise Value Multiples by Sector (US)." Table. New York University Stern School of Business. http://pages.stern.nyu.edu/~adamodar/New_Home_Page/datafile/vebitda.html.

Federal Communications Commission (FCC). 2017. *Twentieth Mobil Wireless Competition Report*. Washington, DC: FCC, FCC-17-126, September 26. Available at https://www.fcc.gov/document/fcc-releases-20th-wireless-competition-report-0.

GSMA. 2016. *Global Mobile Trends*. GSMA Intelligence, October. https://www.gsmaintelligence.com/research/?file=357f1541c77358e61787fac35259dc92&download.

Ovum. 2015. *Telecoms, Media & Entertainment Outlook 2015*. London, UK. http://info.ovum.com/uploads/files/Ovum_Telecoms_Media_and_Entertainment_Outlook_2015.pdf.

Tech4i2, Real Wireless, CONNECT (Trinity College Dublin), and InterDigital. 2016. *Identification and Quantification of Key Socio-Economic Data to Support Strategic Planning for the Introduction of 5G in Europe*. Luxembourg: European Union, 30-CE-0683419/00-45. Available at https://ec.europa.eu/digital-single-market/en/news/5g-deployment-could-bring-millions-jobs-and-billions-euros-benefits-study-finds.

UK National Infrastructure Commission. 2016. Connected Future. London, UK: UK NIC. https://www.gov.uk/government/publications/connected-future.

Chapter 6
Why Key Players Are Enthusiastic

"Worldly wisdom teaches that it is better for reputations to fail conventionally than to succeed unconventionally."

—John Maynard Keynes

The advent of 5G has been marked by unprecedented levels of interest and claim across the globe. Everyone wants to be a leader in 5G. Major manufacturers and operators talk almost daily about how revolutionary 5G will be. Governments sponsor labs and test-beds and push their regulators for early access to spectrum. Alliances are formed between major suppliers and operators. Nobody wants to be left behind.

The first four chapters of this book have shown that there are serious doubts and many unanswered questions about 5G. This chapter discusses why the global industry might be wrong in its enthusiasm for 5G. It is only in the interests of all the key players to talk up 5G's prospects.

The chapter looks at each of the different key segments of the industry and shows why none of them would wish to be critical of 5G.

Academics

Academics love new generations of mobile telephony. It enables them to link their research to a clear goal—to inform the development of standards. It provides opportunities for funding from myriad sources, such as governments, international bodies such as the European Commission, industry, and others. Most of the universities with leading wireless research groups have set up some form of 5G activity, including the 5G Innovation Centre at Surrey University (United Kingdom), the Tactile Internet research at the Dresden Technical University (Germany), and the 5G mmWave center at New York University (United States).

The advent of 5G has been even more interesting than previous generations. As discussed earlier, the trend from 2G through to 4G was predominantly the introduction of better air interfaces in similar frequency bands. But it has been harder to find novel approaches for research. By opening the possibility of mmWave, 5G has enabled new areas of research such as propagation at mmWave, beam forming, massive MIMO antennas, and network design for ultralow latency.

Funding has been at elevated levels, with the European Commission seeing 5G as a major differentiator for Europe and embarking on funding the 5G public private partnership (5GPPP) with €700 million of funding over seven years. This has resulted in many collaborative projects between industry and academia.

Academics are genuinely enthused about 5G and are conducting novel and insightful research. For engineering departments, participation in 5G research is a valuable source of funding, a way to maintain their position as a leading research university, and an opportunity for individuals to make key contributions in fertile new areas of research.

It is clearly not in the interest of most academics to critique the 5G vision; nor are many likely to have the evidence and experience at their disposal to do so.

Equipment Supply Industry

The mobile communications value chain is broad, with many companies having some role within it. The key players are the major manufacturers such as Ericsson, Nokia, Siemens, Huawei, Samsung, and others. Around them sits a huge range of players who provide software, services, test equipment, and much more. It has been the major manufacturers who have been the strongest supporters of 5G and hence the companies on which this section focuses.

Companies like Ericsson and Samsung have been the loudest protagonists for 5G. They frequently issue press releases, predominantly about how fast their latest prototype has transmitted data. They have numerous white papers[1] setting out a bright vision for 5G as a network that will resolve all current problems and provide perfect connectivity (this is discussed in more detail in Chapter 7). They drive the standards process through attendance at 3GPP meetings and invest in academia and other partnerships.

For all manufacturers, the cycle of generations provides a peak in income. A new generation means that they can sell new equipment to network operators. Conversely, as generations become older, upgrades become infrequent. Hence, the industry has become somewhat addicted to the decade-long cycle of investment caused by new generations and is highly reliant on 5G appearing soon to overcome current financial issues. In some cases their future is resting on the success of 5G, and it is no surprise that they have reached new levels of public relations to support its case. With competition intense among them, companies are competing to have the best 5G solution—a solution most simply defined by speed.

---

[1] For example, see University of Surrey, 5G Innovation Centre n.d.

There is an additional dynamic caused by the arrival of new manufacturers such as Huawei and Samsung. While these companies were present during 4G, they tended to follow others rather than take a leading role. With increased confidence they are seeking to change this in 5G. To find a way to gain leadership these companies are being more competitive than they might otherwise have been. This is tending to force the industry toward ever faster performance and away from the quiet discussions that would allow a more pragmatic industry view to emerge.

The manufacturers therefore have a strong interest in seeing 5G succeed and a competitive drive to produce the fastest solution available. It is they, more than any other sector, who are attempting to drive forward the agenda on new generations of cellular.

Mobile Network Operators

The situation of the MNOs was introduced in Chapter 5; they would generally prefer for 5G not to appear since it would result in additional investment, but equally they did not want to be behind their competitors in introducing new technology in case it resulted in a loss of subscribers.

This leaves them with a dilemma. If they were all to critique 5G, pointing out that the costs would lower their profitability and that the services offered were not appealing to subscribers, then this might steer the industry in a more appropriate direction (such as those discussed in Chapter 8). But if one of them were critical when their competitors were supportive, their position might be exploited by their competitors who could show how they were more forward-looking. Also, if regulators felt that these companies were not engaged with future wireless communications systems, then the regulators might be less inclined to listen to them. The safest approach in this case is lukewarm support. Until late 2017 this was exactly what had been seen from most MNOs. They engaged with 5G initiatives and partnered with manufacturers so that MNOs could demonstrate high speeds on their networks and be well placed should 5G prove compelling to consumers. A few, such as Korea Telecom (who was promising to deploy in time for the 2018 Olympics—see Active Telecoms 2016), were vocal supporters.

For MNOs, the risk in being supportive is relatively low during the developmental phase of 5G. Their only commitment is to a few trials and for attendance at some international meetings. They are not being asked to place orders for equipment. The risk is only a lost opportunity should 5G not turn out in a manner that would maximize their profitability.

However, November 2017 saw a sudden change in MNO approach. At a major conference organized by Huawei, both Vodafone and BT were critical of 5G. Vodafone's chief technology officer said that the key benefit of 5G was capacity gains, and that the industry should focus on this rather than new use cases (Davies 2017a). BT's chief executive officer said that he could not find a business case for 5G and, crucially, that he had talked to his peers around the world and that they could not either (Davies 2017b). This appeared to be the beginning of a concerted campaign by MNOs that was needed to avoid leaving any MNO isolated (as mentioned earlier). This action triggered more critical papers (Bicheno 2017) from a number of analysts, who suddenly took a different opinion on 5G.

Even their industry body (the GSMA) started to downplay the 5G hype in 2017. For example, in its *2017 Global Mobile Trends*, the GSMA suggested a slow deployment, saying that:

> Regardless, early 5G deployments will focus on dense city centers using small cells. National rollouts will happen at a slower pace than 4G; by 2025, about 40% of the global population will be covered by 5G. (GSMA 2017, slide 35)

In early 2018, it was unclear how widely the realism around 5G was embraced by MNOs. Despite some MNOs publicly voicing concern, others were still nervous about the implications of doing so. The mood could change slowly or might potentially swing quite quickly in an "emperor has no clothes" type of moment. The next year or so should be interesting.

Government

Governments have generally been strong supporters of 5G. Some believe that their country could be a world leader with 5G and gain business advantage for its companies. Others believe that 5G will deliver services that will improve productivity or otherwise stimulate the economy for their citizens. Many governments wish to ride the wave of the digital future—a future they perceive as having ever faster connectivity.

Governments have taken various actions in support of 5G. Some have sponsored research. Others have engineered test-locations. Some have seen major sporting events, such as the Winter Olympics, as a good venue to showcase their country's industrial prowess. Some regulators have moved relatively quickly to make spectrum available—for example, the United States has been quick to allocate mmWave frequencies.

In large part, governmental aspirations cost little and could potentially deliver political advantage. It would clearly not be in their interest to suggest that there might be a flaw in the 5G vision.

Some supranational bodies have also taken positions on 5G. In particular, the European Commission (EC) has invested heavily and has called for European 5G leadership. They have published a 5G manifesto, "5G Manifesto for timely deployment of 5G in Europe," calling for rapid deployment of 5G into at least one major city in each European country. Observers have also noted that the EC manifesto says nothing about MNO profitability and that the EC's other areas of regulation, such as the removal of roaming fees, work against MNOs by leaving them with insufficient profit to invest in new technology. Further, the panels that the European Commission has assembled to provide expertise on 5G tend to be very heavily composed of the current key players.

Outlandish Claims

As part of the escalating rhetoric, companies and countries have competed to be the first to deploy 5G. Hence, there have been claims such as the 2018 deployment at the Winter Olympics (discussed above). It seems unlikely that the standard will be completed before 2018 and that the equipment will be available by 2019. In practice, the standard decade-long cycle suggests that first equipment availability is more likely to be 2020, with significant rollout by 2022. This aligns with the thoughts of the GSMA reported in Chapter 5. For some elements, such as mmWave, some commentators believe that there is so much research to do that 2025 is a more realistic date for implementation.

Yet many appear to believe, because a major company has announced that it will deploy 5G in 2018, that therefore this must happen. Somehow the cycles of the industry will be speeded up. This has never happened and is unlikely to happen with a standard as uncertain as 5G. Standards take time to develop, and efforts to put pressure on standards bodies tend to backfire as key players stop collaborating and conflicts slow progress. The only way a system can be deployed in 2018 is if it is nonstandard, based on developments within the R&D center of the associated companies. Such an outcome would be harmful to the industry, resulting in nonstandard equipment appearing in different parts of the world; confusing the ecosystem as to what "real" 5G is; and fragmenting the economies of scale. It might result in various players claiming that whatever they deployed in 2018 was 5G.

Such outlandish claims risk making the situation worse by confusing and fragmenting the market.

The Major Players Cannot Be Wrong

The manufacturers and operators are powerful companies. Collectively, they employ millions and account for about $960 billion of annual revenue—about 1.5 percent of the world's GDP. Many believe that if such large companies state they are going to pursue a particular vision or objective, then it is certain to occur. History suggests otherwise. MNOs have in the past declared strong support for:
- Video calling
- Picture messaging
- Location-based services
- Femtocells
- Internet/walled gardens/wireless application protocol
- Widgets/own-brand app stores
- eHealth
- Mobile payment

None of these came to be provided by the MNO. Video calling was delivered over the top (OTT) by Skype. Picture messaging became part of existing apps, with increasingly easy ability to embed pictures in tweets and emails. Location-based services were delivered by Google using data gathered by the phone and then by GPS location. Walled-garden internet was rapidly overtaken by devices like the iPhone, which was able to access mainstream sites. Widgets became redundant with the advent of the app store. MNOs proved unable to make headway into eHealth, and mobile payment was eventually delivered by the banking sector and device manufacturers.

Nevertheless, when there is apparent, widespread support for a particular future, many tend want to bring it about by developing appropriate services and products. Others may find better ways of delivering the concept, as in the example of Skype and video calling.

In this case, it seems likely that MNO support is lukewarm. While MNOs indicate a desire for 5G, they will await its emergence before making decisions as to whether to invest. Major players can be wrong, but it is more a case of their sending out stronger signals of support than they believe internally.

Conclusions

This chapter has shown that it is in the interests of all the key players to be supportive or to be strong promoters of 5G. Academics rely on 5G initiatives for funding. Manufacturers rely on the rollout of 5G to provide a boost in revenues.

Operators fear if they step out of line, they will suffer a competitive disadvantage. Governments see a political benefit in being supportive. It is in nobody's interest to rock the boat. This is in no way intended to be critical—all of these individuals and companies are acting rationally according their apparent best interests.

Perhaps it is worth returning to Keynes's quote. It is safer for all involved to "fail conventionally"—to fall in line with the views of the majority. For example, writing this book involved much personal risk and little personal gain, which may explain why few others follow a similar path.

Chapter 6 References

Active Telecoms. 2016. "Despite Predictions for a 2020 Launch, South Korea Is Striving to Deliver 5G for the 2018 Winter Olympics. Can It Be Done?" *ActiveTelecoms.com*, August 23. http://www.activetelecoms.com/featured/despite-predictions-for-a-2020-launch-south-korea-is-striving-to-deliver-5g-for-the-2018-winter-olympics-can-it-be-done.

Bicheno, Scott. 2017. "There is still a lot of uncertainty around 5G." Telecoms.com, December 6. http://telecoms.com/486636/there-is-still-a-lot-of-uncertainty-around-5g/.

Davies, James. 2017a. "We Need to Stop Talking 5G BS—Vodafone CTO." *telecoms*, November 15. http://telecoms.com/486130/we-need-to-stop-talking-5g-bs-vodafone-cto/.

Davies, James. 2017b. "We're Struggling with the 5G Use Case Right Now—BT CEO." *telecoms*, November 16. http://telecoms.com/486156/were-struggling-with-the-5g-use-case-right-now-bt-ceo/.

GSMA. 2017. *Global Mobile Trends 2017*. GSMA Intelligence, September. https://www.gsma.com/globalmobiletrends/.

"5G Manifesto for timely deployment of 5G in Europe," Brussels, July 7th 2016, Available at: http://telecoms.com/wp-content/blogs.dir/1/files/2016/07/5GManifestofortimelydeploymentof5GinEurope.pdf.

University of Surrey, 5G Innovation Centre. N.d. "Publications: Global System Vendor Community." Website. https://www.surrey.ac.uk/5gic/publications.

Chapter 7
The 5G Vision

Previous generations of mobile systems did not have a vision as such. They were aimed at resolving problems with the prior generation and providing faster connectivity. The discussion in earlier chapters has looked at the issues with providing faster and higher capacity solutions, on the assumption that these will lie at the heart of 5G. But unlike previous generations, 5G proponents have set out more of a vision of new services, some of which revolve around speed and capacity and others around particular capabilities. This chapter introduces the various visions for 5G and considers the extent to which they are practical and economic.

A Collection of Visions

There is no one view on what 5G will offer, but the general expectations are that it will deliver improved user experiences and the potential for new services. To do so, it has been variously suggested that 5G would have to meet one or more of the following criteria:

- Connectivity for multiple devices and services with 100 percent geographic coverage.
- The ability to combine signals from multiple frequency bands (ultrahigh frequency—UHF—up to millimeter wave bands) more flexibly depending on location, time, and application to enable more stable connectivity.
- Low latency to support those services or applications whose requirements cannot be met using existing technologies.
- The bringing together of multiple delivery platforms under a single "umbrella."
- Capacity to support provision of ultrafast broadband services.
- Technology to support a wide mix of different services that require different solutions.

With visions appearing from multiple sources, MNOs came together within the Next Generation Mobile Networks (NGMN) project and summarized their vision for 5G as an:

> End-to-end ecosystem to enable a fully mobile and connected society which empowers value creation towards customers and partners, through existing and emerging use cases, delivered with consistent experience, and enabled by sustainable business models. (NGMN Alliance 2015)

This ill-defined utopia has been reflected in the visions propounded by manufacturers who have suggested performance expectations for 5G of:
- A 1,000 times increase in mobile data volumes.
- A 10 to 100 times increase in connected devices.
- A five times lower latency.
- A 10 to 100 times increase in peak data rate.
- A 10 times battery life extension for low-power devices.
  (METIS and Seventh Framework Programme 2013)

This list is hugely ambitious. Delivering a thousand-fold increase in data volumes would be 50 times better than the best intergenerational volume increase (moving from 2G to 3G) and around 500 times better than that achieved with 4G. Accomplishing this while lowering the latency is even harder—reduced latency tends to reduce efficiency. Visions should be realizable, and this one is clearly not.

The International Telecommunication Union (ITU)—specifically, Working Party 5D (WP-5D)—has developed its own vision of International Mobile Telecommunications (IMT) for 2020 and beyond. This vision highlights access to cloud-based services, augmented reality, high definition video, high speed data access, automation, mission critical communications, and a range of IoT/M2M opportunities. These are shown in Figure 7.1 below.

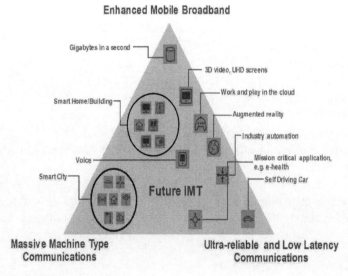

Source: ITU 2015.

Figure 7.1: ITU Vision for 5G

NGMN View of Potential Applications

NGMN (NGMN Alliance 2015) identified 25 different use cases that are grouped under eight different use case families, as illustrated in Figure 7.2 below.

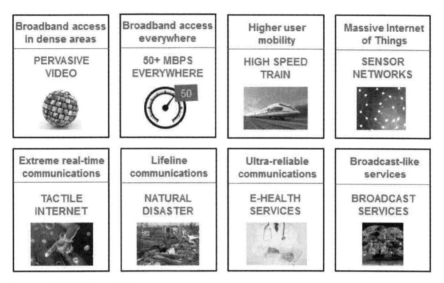

Source: NGMN Alliance 2015.
Figure 7.2: NGMN Usage Cases for 5G

Each of these cases is discussed below.

Broadband Access

The NGMN report has differentiated between broadband access in dense areas and broadband access everywhere. Dense areas are considered to be those where thousands of people live and work or locations such as stadiums.

Broadband access everywhere is intended to provide a minimum experience anywhere and specifically addresses what should be feasible in terms of rural areas and low ARPU areas of the world.

Table 7.1 displays anticipated use cases and challenges.

Table 7.1: Broadband Access Cases

Family	Use cases	Description	Challenges
Dense area broadband access	Pervasive video	High-resolution video used for person-person and person-group communication, regardless of location	Supports high volume of multiple video calls, high data rates, low latency
	Smart office	Devices in office are wirelessly connected to the cloud and use high bandwidths.	High volume of traffic; low latency in some cases
	Operator cloud services	Addition of value-added services and more customization for each user	Consistent, fast, and reliable networks; seamless interworking across networks
	HD video/photo sharing at stadium and other venues (indoor and outdoor)	High connection density and temporary use for an event; sharing of live video or photos on social networks	High volume of connections, high data rates, low latency
Broadband access everywhere	50+ Mbps everywhere	Provision of a minimum of 50 Mbps data rate per user everywhere	Data rates to be supported at cell edges with effective handover
	Ultra low-cost networks	Lower cost provision of basic services	Flexibility to be able to deliver solutions for dense and scarcely populated areas

Broadband access in dense areas is already provided by 4G. The challenge as outlined here is to deliver sufficient capacity in locations such as within an office or stadium to enable applications. As discussed in Chapter 3, it is unclear whether this challenge actually exists. If data requirements do reach a plateau in mid-2025, then it will generally be possible to meet capacity requirements with existing 4G systems that are augmented with additional spectrum in bands such as 3.5 GHz. Data demands in office and other indoor locations are already being met with Wi-Fi, and increasingly dense Wi-Fi deployments are providing capacity in stadiums and other similar venues. If the need did exist, then the only new tool that appears to be at the disposal of 5G is the use of mmWave deployments. Those would certainly meet the capacity requirements but at a cost that looks highly uneconomic.

Broadband access everywhere is a worthy aspiration and one that is examined in more detail in Chapter 8. It requires enhanced network coverage. MNOs stopped enhancing coverage some years ago when they reached the point of economic viability. Changing this would require a technology that could deliver coverage at a lower cost. There appears to be no solutions within 5G that would provide lower cost than those available within 4G. With core frequency bands for 5G at 3.4 GHz, coverage will be very limited compared to 4G deployments in sub-1 GHz bands.

To sum up, dense area broadband access is already provided with 4G and Wi-Fi, and capacity needs will plateau. Broadband everywhere requires lower cost solutions, whereas 5G proposals all appear to be higher cost.[1]

Higher User Mobility

The expectation is that there will be a growing demand for mobile services while on the move. Use cases are set out in Table 7.2.

---

[1] A report for the National Infrastructure Commission (Oughton and Frias 2016) suggested that for the United Kingdom, meeting the 5G aspiration of 50 Mbps everywhere would cost £71 billion (total cost of ownership), whereas achieving 10 Mbps everywhere would cost £20 billion. Neither are judged to be affordable.

Table 7.2: High Mobility Cases

Use cases	Description	Challenges
High speed train	For passenger access to high quality mobile internet while traveling (e.g., access to videos, movies, the internet, the cloud, office systems, and video conferencing)	A potential for high density of users and need for low latency and robust communications when traveling at high speeds
Remote computing	For remote computing to be available while on the move	Low latency and robust communications
Moving hot spots	For events (similar to mass events at stadiums) that may be over a much wider geographic area	Need to provide additional capacity in real time to meet unplanned requirements
3D connectivity	For provision of commercial services on aircraft similar to those on the ground	Need to provide services above the ground

Many are familiar with the problems of connectivity while traveling. Trains pass through tunnels where there is no signal. Cars drive through areas of weak connectivity, and only a minority of planes are equipped with Wi-Fi. These are predominantly problems of network coverage. If 4G or Wi-Fi were provided, it would deliver the capabilities needed, with 4G being able to accommodate the handoff requests that occur as groups of passengers pass through cells. There are generally good reasons why coverage is not currently deployed. These include the difficulty in gaining access to train tracks or the uneconomic cost of covering all roads. These issues are explored in more detail in Chapter 8. As with broadband provision (above), there is nothing in the 5G road map that will change these difficulties. Some aspects—such as larger, more complex antennas—may actually add to the difficulties.

There are problems here that deserve to be solved, but they do not need a new generation of mobile technology.

Massive Internet of Things

The new 5G technology is seen as meeting the needs of an expected massive number of devices (sensors, actuators, cameras) that will be deployed to monitor or measure a wide range of different attributes. These specific use cases are identified in Table 7.3.

Table 7.3: Internet of Things Connectivity Requirements

Use Cases	Description	Challenges
Smart wearables	For integrated sensors that can be used to measure environmental and health attributes, such as blood pressure, temperature, and heart rate	Management of the devices and data
Sensor networks	To be used for metering; environmental management (e.g., pollution, noise measurements); traffic control (e.g., traffic light management)	Device costs, battery life, high density, wide range of devices, transmission distances
Mobile video surveillance	To be used for surveillance on moving transport (aircraft, cars, drones); by security personnel to monitor events, buildings, etc.	Need for reliable and secure networks

Source: NGMN Alliance 2015.

The concept of the IoT was introduced in Chapter 3. IoT was shown that to be a highly valuable service, but one with relatively minimal data requirements. The question now is whether 5G is needed for IoT to materialize.

There are many proposed solutions for IoT connectivity. They divide into cellular-based solutions and into noncellular solutions running in unlicensed frequency bands (and hence often termed "unlicensed solutions"). Within 4G there are two main solutions—LTE-Machine (LTE-M) and narrowband IoT (NB-IoT). In 2017, both of these were in trials in multiple countries, and many MNOs had committed to roll them out across their network. Unlicensed solutions include Sigfox, LoRa, Weightless, and Ingenu and are being deployed to varying extents around the world.

It is hard to see a role for 5G. Solutions such as NB-IoT are well specified and should meet the foreseeable needs of IoT. LTE and 4G offer sufficient capacity and coverage. Deploying another IoT solution within 5G would add further expense for MNOs, requiring them to run multiple networks in parallel and threatening to confuse the marketplace. There are no unmet needs currently identified nor any new technology that would allow 5G to offer a superior solution to 4G. In a major 5G debate (Cambridge Wireless 2016a,b), the knowledgeable audience were unanimous in their view that 5G would not introduce a new IoT solution but that it would encapsulate the existing 4G solutions. This was also the view that the 5G standards bodies appeared to be reaching in 2017. Of course, if 5G becomes just an encapsulation of existing technologies in all areas, then there is no need for it at all, since these are already broadly encapsulated within 4G.

This is not to claim that all is perfect in the IoT connectivity market. There are too many competing solutions that are causing market fragmentation and confusing potential users. Some solutions are proprietary and are unlikely to succeed in a world where the only successful wireless technologies are open standards. The business case for MNOs is still uncertain, and for unlicensed operators there are concerns about interference and longevity. But these are not problems that a new 5G technology would solve; instead, they require market consolidation, maturity of offerings, and stability.

In summary, IoT is an important area, but one well served by existing technology.

Extreme Real-Time Communications

This is intended to cover those uses that require real-time interaction. This could be for control of electricity smart grid networks, autonomous cars, or robotic control and interaction (e.g., remote medical care). One specific use case identified is tactile internet, where real and virtual objects are wirelessly controlled by humans.

The tactile internet is based on an observation that when humans interact directly with objects by touch through their fingers, the feedback they get in terms of pressure as they touch objects happens within 1 millisecond (ms) (although some suggest this takes 10 ms). If this feedback takes longer then it becomes difficult to control complex machines by touch—the effect is akin to that of an echo in speech. Current 4G systems could achieve a delay, or latency, of around 10 ms, although 50 ms is more likely in practice in a loaded network. Hence, 4G is an order of magnitude away from delivering tactile capabilities. Some believe that if the latency in 5G could be reduced to 1 ms, it would enable a suite of new applications such as remote surgery.

The extent of these applications is little understood. Remote surgery is clearly a very niche service, only needed by tens of people in a country. There may be other remote-control applications with more widespread demand, but these have yet to be identified. Applications such as VR would also benefit from low latency—in the case of VR, less than 10 ms would be ideal, although 20 ms may be sufficient for most.

Almost all the services suggested are likely to be used indoors. Remote surgery is definitely an indoor occupation, as is VR.[2] In that case, these applications could be connected via a wire (e.g., to a VR headset) or via short-range wireless such as Wi-Fi. This suggests that if low-latency radios are to be designed, it would be more helpful for them to be modes within short-range communications than in cellular. It also suggests that indoor connectivity can be used to test the demand for low-latency services. Should it become clear that there are applications that (1) large numbers of users wish to take outdoors and (2) users will pay substantially more for, then the business case for an mmWave deployment or similar feature might become feasible.

Delivering low latency is challenging. It would require radical redesign of many of the concepts in mobile radio systems. Further, it tends to work against spectrum efficiency as, for example, large blocks of data can be encoded more efficiently than smaller ones. Hence, there might need to be a trade-off between efficiency and latency. If that were the case, the economics of network operation would suggest that unless the low-latency services were highly valuable, efficiency would always come first.

In summary, while the tactile internet is an interesting concept, it is very difficult to deliver, likely to compromise other network functions, and has uncertain revenue. Unless clear demand emerges, it is hard to see MNOs building it into their networks.

Lifeline Communication

It is anticipated that further applications will emerge for public safety and emergency services. These applications will typically require a high level of availability, need long battery life of user terminals, and cater for surges in traffic. One specific use identified is communications in the event of natural disasters.

Emergency service communications has been the subject of much debate, with some now tending toward the use of 4G. Recent releases of the 4G specification have added features that these user groups need. If these users deploy a 4G system in the coming years, it seems inconceivable that they would rapidly replace it with 5G. Such user groups tend to keep their systems for some time—for example, railway communication is still based on a variant of the GSM system.

---

2 Augmented reality (AR) is an outdoor application that requires cellular connectivity. An early example—Pokémon Go—became extremely popular in 2016. However, it only added around 0.1 percent to total network traffic and did not require low latency.

Emergency services will adopt 4G for their needs, and there is nothing in the 5G specifications at present that would suggest they should wait for 5G instead.

Ultrareliable Communications

The NGMN white paper notes that the vision of 2020 and beyond not only covers automotive, health, and assisted living but also covers industries such as manufacturing through agriculture that require reliable machine type communications. Other applications include those where there is a need for remote operation and control; in such instances, there is often a requirement for low latency (e.g., smart grids, automated traffic control and driving). There is often no (or limited) need for mobility. Use cases are given in Table 7.4.

Table 7.4: Communications Requiring High Reliability

Use Cases	Description	Challenges
Automated traffic control and driving	Advanced safety applications to mitigate road accidents and improved traffic efficiency and support mobility of emergency vehicles; requires vehicle-to-vehicle and user (pedestrian and cyclist) to vehicle communications	Low latency for warnings as well as higher data rates to share video information; high reliability for many cases
Collaborative robots	Automation in (for example) manufacturing	Very low latency and high reliability
eHealth	Remote health monitoring and remote treatment (e.g., ECG, pulse, glucose levels)	Urgency/priority, out of coverage, security/authentication
Remote object manipulation	Remote surgery in mobile instances (e.g., in ambulances or disaster situations); control of remote devices used, e.g., to provide information on hazardous locations/incidents.	Latency, reliability, security
3D connectivity	Connection with devices such as drones	Reliability
Public safety	The means to send video, pictures, data information from any location	Priority, reliability

Source: NGMN Alliance 2015.

While this list of applications may not be comprehensive, it is worth discussing each in turn to understand the issues and benefits of a potential 5G solution.

Automotive. Vehicles work well today with limited, if any, connectivity. The key development that may change this is autonomous vehicles. Some have suggested that effective autonomous operation will require very high reliability and very low latency connectivity so that the car is always under the control of a central network.

But a moment's thought shows that this cannot be so. If an autonomous vehicle required such connectivity, then it would have to stop whenever the connectivity was not available. Given the parts of road network that do not have coverage, this would mean that autonomous cars would make it a few miles from the dealership before halting until network coverage was expanded, which could be years. This is clearly ludicrous. Autonomous vehicles are so termed because they can operate autonomously—without any network connectivity. This is exactly how early solutions from Google and Tesla work. When there is connectivity, the vehicle may download useful information such as traffic conditions, but most of the time it controls its operation locally. Centralizing control would be expensive, prone to delay, and extraordinarily complicated (Bubley 2016).[3]

Vehicles could usefully be connected. Telemetry data from the engine and other parts of the vehicle could help with maintenance. This can be provided with emerging IoT solutions. Passengers might enjoy broadband connectivity for entertainment—this can also be provided with 4G. Cars may wish to communicate with their neighbors (for example, to indicate abrupt braking)—this can be done with simple short-range direct radio systems. Of course, if super high speed connectivity was widely available alongside roads, passengers might find it attractive to stream video, but the cost of such a provision seems highly likely to outweigh the value that passengers would place on it. Instead, they would download video in advance of a journey, as they do now.

Given the key part that many believe autonomous vehicles will play in 5G, a more detailed discussion of autonomous vehicles is provided in this chapter's appendix.

Robots. Robots are generally static devices on a production line that is linked with a cabled solution. If a wireless link is needed, it is provided using unlicensed systems within the factory, such as Bluetooth or Wi-Fi. It is unlikely that 5G will

---

[3] Analysys Mason published an insightful paper on the needs of the autonomous car (or lack of them); see Rebbeck 2017.

have good indoor coverage, and the costs of using it compared to a self-provided system will be high. There is no role for 5G here.

eHealth. Remote monitoring of a patient's vital signs could be of immense value to some, such as those suffering from chronic conditions. It is generally achieved today with a monitor communicating via Bluetooth to a phone which then relays data as needed. In the future, it may be simpler to connect directly using IoT devices. The point of concern is whether urgent messages can be reliably received. If such a message is blocked due to network congestion or lack of coverage, the effect could be life threatening. The 4G network has the ability to prioritize traffic. Adding a further network to the mix would seem to add little. Indeed, if the expenditure displaces those monies targeted for improved coverage, the effect of 5G could be detrimental.

Drones. Drones tend to be controlled via short-range wireless connectivity. Sometimes this is proprietary; other times, solutions such as Wi-Fi are adopted. These solutions work well until the drone passes out of sight or out of range. However, most drones are restricted by regulation to remain within sight of the operator. Further, the market for drones is relatively small, and the idea of deploying a completely new network to service it would be uneconomical.

Remote Surgery. This was covered earlier in this chapter. Broadly speaking, it is an indoor activity where short-range wireless solutions would be used.

In summary, while there are some valuable applications here, they can be well served with existing solutions. There is nothing of sufficient novelty, scale, and value to make the deployment of an expensive 5G solution worthwhile.

Broadcast-Like Services

Some believe that there will be a move away from broadcast TV toward both real-time and non-real-time services that distribute content via the cellular downlink. The content can also provide a feedback channel (uplink) for interactive services. Local services may provide offers and information from local companies, and regional services may provide traffic congestion information.

The problem is that this has been tried before—with failure. Within 4G there is a well-specified broadcast mode termed evolved multimedia broadcast multicast service (eMBMS). This has been through extensive trials by MNOs and examined as part of a broader European broadcast strategy by the European Commission

(Lewin et al. 2014). The trials proved only to have limited value, and the European Commission concluded that there was no viable business case for widespread broadcast via cellular networks. There is nothing new in the proposed 5G specifications that would change these conclusions. Equally, there is nothing to prevent a new business case emerging that could make use of the capabilities already available in 4G.

Summary of NGMN Requirements

The various requirements set out by NGMN fall mostly into three categories:
1. Better coverage.
2. New services.
3. Consolidation of functions of other networks.

Better coverage appears best delivered through expansion of 4G rather than via 5G. There are no technologies within 5G that deliver greater range, and its deployment in the higher frequency bands makes it more suited to urban areas.

New services such as the tactile internet offer some promise, but the economics appear very challenging. Without the prospect of significantly increased revenue, it is hard to see how MNOs could make the business case for their deployment.

Consolidation of other functions such as IoT and broadcast appear unnecessary. The advantages of consolidation are minimal and 5G does not bring any new concepts or technology.

Despite the breadth of the user cases (or perhaps because of it), there is little that MNOs could use to convince investors to provide them with the capital needed for 5G deployment. By providing such a broad "wish list," it has made focusing on 5G difficult; the result is unconvincing.

Conclusions

The 5G community cannot be accused of being short of visions—quite the converse. Visions range from metrics for the radio system to a wide breadth of usage cases. It seems that 5G is intended to solve all the problems of the mobile community and provide a utopian solution where all have perfect communications that meet every need that they might have. Compare this with previous generations, where the visions have been much more restricted, such as improving capacity or providing a specific data rate.

But these visions are too utopian. Fully achieving them would require astonishing breakthroughs in radio technology and would require subscribers to be prepared to significantly increase their spending. Both are heroic assumptions. In practice, most visions can be adequately achieved with existing technology, such as evolved 4G, evolving Wi-Fi, and emerging IoT technologies. Those visions that are more difficult to deliver can be readily critiqued; simply claiming that optimism is needed to resolve the problems is insufficient.

The future is uncertain, and requirements or services may well emerge that result in a different industry than seen today. But until that happens, 5G investment remains unattractive.

Chapter 7 References

Bubley, Dean. 2016. 5G vs. AI: "Do Smart Devices Still Need the Fastest Networks?" *LinkedIn*, November 10. https://www.linkedin.com/pulse/5g-vs-ai-do-smart-devices-still-need-fastest-networks-dean-bubley?trk=prof-post.

Cambridge Wireless. 2016a. "Debate 1: Technology Readiness." The CW 5G Debate in association with the NIC held at the Shard, London, United Kingdom, October 24. https://www.cambridgewireless.co.uk/media/uploads/resources/Debates/24.10.16/5G_Debate1_24.10.16_FinalTranscript.pdf.

Cambridge Wireless. 2016b. "Debate 2: Business Cases." The CW 5G Debate in association with the NIC held at BT Tower, London, United Kingdom, November 8. https://www.cambridgewireless.co.uk/media/uploads/resources/Debates/08.11.16/5G_Debate2_08.11.16_FinalTranscript.pdf.

ITU Recommendation M.2083: IMT Vision - "Framework and overall objectives of the future development of IMT for 2020 and beyond" https://www.itu.int/rec/R-REC-M.2083Lewin, D., P. Marks, Y.S. Chan, W. Webb, C. Chatzicharalampous, and T. Jacks. 2014. *Challenges and Opportunities of Broadcast-Broadband Convergence and Its Impact on Spectrum and Network Use*. Luxembourg: European Commission, 30-CE-0607487/00-85. http://www.plumconsulting.co.uk/pdfs/Plum_Dec2014_Broadcast-broadband_convergence_and_impact_on_spectrum_and_network_use.pdf.

METIS and Seventh Framework Programme. 2013. Deliverable D1.1: Scenarios, Requirements and KPIs for 5G Mobile and Wireless System. April 29, ICT-317669-METIS/D1.1. https://cordis.europa.eu/docs/projects/cnect/9/317669/080/deliverables/001-METISD11v1pdf.pdf.

Next Generation Mobile Networks (NGMN) Alliance. 2015. *NGMN 5G White Paper*. Frankfurt, Germany: NGMN Ltd., February. https://www.ngmn.org/fileadmin/ngmn/content/downloads/Technical/2015/NGMN_5G_White_Paper_V1_0.pdf.

Oughton, Edward J., and Zoraida Frias. 2016. *Exploring the Cost, Coverage and Rollout Implications of 5G in Britain*. London, UK: National Infrastructure Commission. https://www.nic.org.uk/publications/exploring-cost-coverage-rollout-implications-5g-britain-oughton-frias-report-nic/.

Rebbeck, Tom. 2017. *Autonomous Vehicles: Exploring the Opportunities for Operators*. London, UK: Analysys Mason. http://www.analysysmason.com/Research/Content/Short-reports/autonomous-vehicles-operators-RDME0/.

Chapter 7 Appendix: Autonomous Vehicles in More Detail

For some time, autonomous cars such as those from Google and Tesla have been driving themselves around California without much connectivity. But now we are told that ubiquitous connectivity delivering an ultrahigh data rate and millisecond latency is critical to the future of the autonomous car. What has changed to require this? Or, is this not actually a requirement at all, but rather a 5G industry seeking a use case to push 5G?

A car might need many different forms of connectivity. A telemetry connection might enable uploading of engine and vehicle information and the download of software updates. This would typically only need to happen infrequently, such as daily, and could be achieved using cellular and even satellite for the downlink. Passengers might require connectivity for entertainment and work purposes; this is delivered today either directly to the devices using 4G or via a Wi-Fi repeater in the car, often working from a 4G connection on the roof. Finally, autonomous vehicles might need map updates, congestion information, and perhaps control information. Telemetry and passenger communications are already achieved with existing connectivity; it is the autonomous operation that is considered in this appendix.

Two extreme views on autonomous operation could be imagined. At the one extreme, autonomy really does mean autonomy. The car is on its own with no connectivity, only the preloaded maps, and it navigates itself around, using on-board sensors to make decisions about accelerating, braking, and so on. This is termed "truly autonomous." At the other extreme there is no autonomy and the car is under the complete control of a network directing its every move. This is termed "network control." Current autonomous cars are closest to truly autonomous, whereas the future envisioned by the 5G community is network control.

In both scenarios, there might also need to be vehicle-to-vehicle (V2V) communications. The most obvious use for this is when a vehicle is breaking heavily to signal to the one behind it to warn that vehicle to slow down too. That vehicle can then signal back to the one behind it as needed, and so on. This does not need network control and can be enabled with a simple wireless short-range link. Indeed, an even simpler solution would be for brake lights to glow more intensely when the braking is harder. Humans driving nonautonomous cars could see this and take action, and autonomous cars could use light detectors to pick up the intensity. Light has the advantage of only being seen by the vehicle behind, preventing other vehicles in the vicinity (such as those on the opposite lane) braking unnecessarily. V2V could also be used for warning messages, such as an icy patch, and this could be done, for example, via a particular modulation of the brake lights.

V2V is a useful feature that will happen independently of whether there is network connectivity or not, and not by relying on any telecommunications standard such as 5G.

The truly autonomous car is clearly possible as it is currently being demonstrated in trials over millions of miles of driving. Is the network control car any better? Full network control is clearly not viable. This would require connectivity across every mile of every road. Without it, the car would need to come to a halt and await human intervention. Such connectivity does not exist now and is unlikely to do so in a 5G era, where the higher radio frequencies deployed will result in much less coverage. Indeed, low-latency 5G connectivity might only be available in city centers. Hence, all cars need to be truly autonomous.

This leads to a refinement of this question: is a car with occasional network control better than one without network control? "Better" in this context means bringing benefits that outweigh the costs to the vehicle's owner. Proponents of the network control car mainly argue that it is safer, with a secondary argument that it might allow greater road capacity through platooning and other congestion management techniques. Opponents note that the connectivity needed will be extremely expensive, and the central computer system needed to control a nation's cars is massively complex, is liable to have bugs, and will need funding and regulatory approval. Low-latency connectivity then will be only available in dense areas and expensive as well. Let us now look at safety and road capacity in this context.

Since autonomous vehicles will spend most of their time being truly autonomous (because low-latency connectivity will not be available on much of their route), they will clearly need to be very safe. Google and others have already demonstrated levels of safety well beyond those of human drivers. This will likely only get better both as the algorithms improve and as the percentage of autonomous cars goes up. When collisions do occur, they tend to be at low speed. How might network connectivity help? Imagine a case where a child runs out into the road after a ball. A centralized network will not know about this, so the first car will need to detect the child using its sensors and take evasive action. It will likely warn the cars behind it with V2V so they will know about it. But a car that is coming around the corner and is too far back to see the V2V signal might be surprised on encountering stationary vehicles. In principle, a network could warn it about this. In practice, a car that ensured its braking distance was always less than its forward visibility would not need such a warning. Of course, there are many more scenarios to consider, but it is hard to see network connectivity as providing a material improvement in safety and certainly not one for which users will pay hundreds of dollars a year.

Road capacity is important, and platooning can help. Here, a car might need a message to join a platoon that is perhaps a little distance ahead, so it needs to speed up until it reaches it. Such a message can be sent over existing connectivity. The car can then join the platoon and fine-tune its position using its sensors. If the lead platoon vehicle needs to brake suddenly, then V2V can propagate this information backward through the platoon probably more quickly than even a low-latency network could.

It is hard to see how low-latency 5G connectivity could add sufficient value to be worth the cost of deployment. In any case, enough autonomous cars to make it all worthwhile—perhaps 20 percent of the car population—are very unlikely to appear within the next decade, making this more of a 6G problem than a 5G problem.

Precision network connectivity of cars has very little benefit, will be very expensive, and will only be available in dense urban areas. In any case, the volume of autonomous cars needed to make the economics work will not exist during the lifetime of a 5G network. This suggests that the autonomous car industry does not have any compelling need for 5G for which it would pay significantly. It is more likely that the 5G industry needs the autonomous car to adopt 5G in order to provide a use case and business case.

Chapter 8
Alternative Futures

Previous chapters have concentrated on showing the flaws in the vision for 5G. They have suggested that users do not value faster speed and that the need for capacity growth will come to an end shortly after the introduction of 5G. However, in Chapter 7 a number of visions were presented for 5G which have value; this includes ideas such as enhanced coverage. Chapter 8 discusses those elements of the 5G vision that are valued by consumers and shows how they might come about.

Why Consistency Is More Important than Speed

Chapter 3 shows that if the internet servers were always responsive, and if mobile users always had a good signal level in uncongested cells, then speed would be more than adequate for all of the applications commonly in use today. The problem is that all of these conditions are rarely met. The situation is similar to road infrastructure—all people would like quicker journeys, but the limiting factor is not the top speed of cars but road capacity. That is why ad hoc surveys and anecdotal evidence suggests that, for many, speed has reached the point where further gains are of limited value; what is becoming much more important is consistency.[1] Most people would rather have satisfactory data rates available everywhere than they would have blinding fast rates in some places and a lack of any connectivity in others. Likewise, for most vertical applications—for example, constant connectivity even at relatively low rates—would be more helpful for autonomous vehicles than erratically available high data rates.

Concentrating on consistency (now that there is a basic sufficiency of data rates) is also more likely to improve productivity and social value—certainty of having a connection that would enable new methods of business, better responsiveness, etc. Conversely, speeds above 10 Mbps (to the home) are currently almost entirely used for entertainment, which enhances pleasure but not productivity. With governments looking to improve productivity, global competitiveness, and more, a focus on consistency rather than speed appears appropriate.

---

[1] This begs the question as to why some MNOs still advertise the high speeds their networks can achieve. This appears to be "bragging rights"—using an attribute few care about directly to demonstrate the strength of the network. It is akin to car makers promoting high-performance models that few will buy.

How to Deliver Consistency

True consistency is hard to deliver and hard to measure—there will always be a basement or remote area that does not have coverage. Instead of a generic focus on consistency, it is better to look at those areas where coverage or capacity is most obviously problematic. These include:
- Transport: specifically, trains and, to a lesser extent, buses.
- Rural areas.
- Buildings, including homes, offices, and public buildings.
- Very dense areas, such as major train stations.

Each of these is considered below.

Trains

MNOs have been trying to provide good coverage within trains for many years, with variable success. Coverage problems tend to occur when any of the following happen:
- Railway cuttings or tunnels that block the radio signals from getting to the outside of the train.
- Metallized windows on passenger cars, which create a block between the outside and the inside.
- High speed trains, where handovers can occur so frequently that handover traffic dominates.
- Trains operating in dense urban areas, where the demand for capacity is very high.

The best solution to most of these is a Wi-Fi repeater within the train—and indeed this is increasingly being widely deployed. A repeater overcomes the isolation problem caused by metallized windows—indeed, it benefits from this isolation, as it reduces any external interference which helps in areas with high demand outside of the train. It also solves the handover problem—the devices inside the passenger cars stay registered onto the internal Wi-Fi access point. It can also help somewhat with the problem of railway cuttings and tunnels by using an external antenna mounted on the roof of the train, with much better performance than handsets within the train.

The repeater could also transmit cellular signals alongside Wi-Fi signals. However, this tends to be problematic because cellular transmissions have to be on licensed frequencies owned by the MNOs. Gaining their approval and then selecting frequencies that do not cause interference to their external network is difficult. The repeater becomes much more complex, as it has to cover multiple bands. Finally, for most users, data connectivity is more important than voice because the users can then browse, receive emails, and make calls using voice-over-Wi-Fi solutions such as Skype and WhatsApp. The only problem is not being able to receiving incoming calls via the cellular network.

However, the repeater transfers the coverage problem to the backhaul connection between the passenger car and the network. With many tens of users in a passenger car, for all to be able to use laptops or tablets, total data rate requirements within a packed passenger car could potentially exceed 100 Mbps at peak times. That is beyond the capacity of most existing backhaul solutions. Backhaul to a railway car could currently make use of both cellular connectivity and satellite connectivity. The best solutions use both—relying on satellite when outside of cellular coverage. But neither satellite nor cellular can provide speeds of 100 Mbps; instead, around 10 Mbps is more likely, and both fail in tunnels.

So, to properly resolve coverage on trains, a twofold approach is needed:
- Wi-Fi repeaters installed in all of a train's passenger cars.
- Better backhaul coverage to passenger cars.

The first is an economic and logistical problem. There is a cost involved with the installation, and it can only be performed when a passenger car can be routed to a maintenance depot. This can only be solved by appropriate economic incentives (such as a requirement on the owners of the passenger cars to provide repeaters) and by allowing sufficient time for passenger cars to rotate through maintenance depots or be replaced with newer rolling stock.

The second problem requires base stations for the delivery of backhaul connectivity. These should be mounted alongside the track, where they can provide good coverage along the length of the line, and especially into cuttings. It may require specialized solutions, such as leaky-feeder cable installation in tunnels. This has previously been problematic because each MNO needs access and because restrictions on trackside working and deployment of equipment have made such base stations expensive and, in some cases, intractably difficult to deploy. However, railway network operators generally manage to deploy their own base stations to provide operational communications. The problem will be eased by restricting access to a single backhaul provider rather than to all MNOs. The equipment deployed would be configured to provide optimal backhaul connectivity (rather than direct to the phone connectivity), perhaps operating in the mmWave region where there is ample bandwidth.

These are predominantly logistical problems, requiring legislation and incentives on various players in the railway industry to resolve. They could materially improve train communications to deliver a better connectivity than is possible today.

Rural Areas

Covering rural areas is predominantly a matter of economics. It could be achieved with widespread deployment of cellular masts (or towers), but these masts would generate less revenue than they would cost to deploy and maintain. Hence, few (if any) mobile operators would voluntarily deploy. Getting better coverage could be achieved by:
- Appropriate financial incentives, such as a subsidy from the government in return for achieving certain coverage objectives.
- Technology that enables a greater range from a base station, thus requiring fewer base stations and so making the coverage more economical.

The first tends to happen indirectly using coverage obligations in spectrum licenses. However, a better approach might be to encourage MNOs and others to bid to deliver the required coverage. The government would then select the best bid and pay the winner to deploy their solution. The network they deploy could then be shared among all MNOs, such that all subscribers gain coverage at the lowest cost to the government.

The concept of subsidizing rural coverage is far from a new idea. For example, in Australia, the federal government has a Mobile Black Spot Program that is intended to improve mobile coverage and competition in regional and rural Australia through subsidizing the cost of building new base stations in areas without coverage. The government committed AUS$100 million in Round 1, which will deliver nearly 500 new and upgraded mobile base stations across Australia (Australian Government, Department of Communications and the Arts n.d.).

Standards bodies have not tended to focus on technology that extends range, as this is generally at the expense of higher data rates. The classic solution, used extensively in many IoT systems, is to use data spreading (known as direct sequence spread spectrum, or DSSS) to increase the range at the expense of the data rate. This is precisely the solution used by GPS satellites to enable a low-power transmission from orbit to be received by small devices. Adding a DSSS mode into the standards would give operators flexibility to trade off data rates against range when it is appropriate to do so, facilitating rural coverage. Unfortunately, such a mode does not currently appear to be on the agenda of the key standards bodies.

In the Home

For most, data coverage in the home is provided via self-deployed Wi-Fi, generally giving excellent data rates as long as the home broadband connection is acceptable, there is not significant Wi-Fi interference, and the signal level throughout the home is strong. Interference can often be addressed by changing channel and poor signal levels by using repeaters or better siting of the access point.

The remaining issues are then cellular coverage and possibly coverage for visitors.

Cellular coverage can be important, particularly in receiving incoming calls. Outgoing calls can be made from the home cordless phone or using Wi-Fi calling apps. Various attempts have been made by MNOs in the past to get in-home coverage using femtocells, but these have mostly failed because:
- Home owners do not want an extra box in the home.
- The solution is tied to one MNO (unless multiple boxes are installed), making a switch among MNOs harder. This may not be suitable for all members of the family.
- Integrating the home femtocell into the MNO's network can be complex and expensive.

As Wi-Fi continues to gain traction, it seems unlikely that femtocells will see a resurgence. Instead, ways around poor cellular coverage by using Wi-Fi will be developed for the home.

Wi-Fi coverage for visitors can be achieved just by telling the visitor the password. This is workable but somewhat clunky; a more automated process could be envisioned and might be part of a broader solution to automate the process of signing into Wi-Fi access points. This is discussed in more detail in the next section.

In the Office

To a fair degree, office issues involve the same set of issues as in the home. Wi-Fi provides a good solution, but cellular coverage can be poor. Femtocells and small cells have not proven widely popular, and that seems unlikely to change. Using the same set of solutions as the home to provide Wi-Fi calling and a simplified way to gain passwords would resolve most issues.

Public Buildings

Technically, public buildings are not materially different from office buildings (although some buildings, such as museums, can be larger and more challenging to cover). Hence, as with the home and the office, the same solutions apply regarding Wi-Fi. Administratively, this requires the deployment of Wi-Fi[2] and a mechanism to enable easy access. If government did deploy a universal password solution for public buildings, this might be of value in delivering universal password solutions more widely—for example, the same solution could be adopted for homes and offices. Alternatively, government could make use of developing solutions in the private sector.

Dense Areas

Areas of very high user density—such as major train stations and stadiums—present particular problems. Cellular solutions struggle to cope with the need for extremely small cells in often a very open environment, where there is little to prevent interference from one cell to another.

In stadiums, there are specific Wi-Fi solutions whereby access points are deployed across the inside of the roof, providing targeted downward pointing beams that might illuminate only 10 or 20 seats. Similar solutions could be deployed for cellular, but again it is difficult to deploy one solution per operator, and the building owner may prefer to deploy a self-owned and self-operated solution rather than negotiate with the MNOs.

---

[2] The European Commission has recently proposed to make funding available to assist in such deployments. See Procedure 2016/0287/COD, amending Regulations (EU) No. 1316/2013 and (EU) No. 283/2014 as regards the promotion of internet connectivity in local communities.

Similar solutions could be envisaged in train stations. At present, most Wi-Fi in these venues is provided by shop owners in an ad hoc manner that causes poor coverage in some areas and interference in others. Centralizing the planning and deployment of Wi-Fi would dramatically improve the situation. This would require agreement from shop owners, some of who might deploy specific solutions as part of their franchise (e.g., access in Starbucks). As with railway coverage, it might take direct government intervention to bring about an improvement in major train stations. In areas such as malls, there may be sufficient commercial self-interest from the mall owner to have a centralized deployment.

Summary

Across the various solutions, there have been a number of common threads—namely:
- Intervention from government in aspects such as train stations—Wi-Fi in major stations and trackside coverage to force through change—and in awarding contracts for rural coverage.
- The sharing of infrastructure among all MNOs in rural areas and possibly in other places.
- The addition of a DSSS mode in cellular to enable greater range for rural coverage.
- The ability for incoming cell phone calls to be rerouted across Wi-Fi, such that if there is no cellular connectivity, people are still in contact.
- The ability for devices to be sent information on a service set identifier (SSID) and password rather than users having to ask for passwords and manually enter them. This could be generic (along the lines of the UK's BT Wi-Fi [formerly BT OpenZone], where any BT Wi-Fi customer can use the Wi-Fi router of any other customer), or it could be based on various criteria (e.g., allowing friends on Facebook access to the password, linking a hotel booking through a browser with a download of the Wi-Fi details, etc.).

Each of these is considered in subsequent sections.

A "Wi-Fi First" World

Previous calls for enhanced coverage have mostly focused on cellular, and previous efforts to provide widespread Wi-Fi "municipal" coverage have generally been seen as a failure. The steps set out in the previous section would move Wi-Fi back to center-stage in the world of communications. Is this plausible, and have lessons been learned from previous attempts to deploy widespread Wi-Fi?

It is worth recalling that we already live in a Wi-Fi-first world. Well over 50 percent of the traffic from our mobile phones flows over Wi-Fi, and typically 100 percent of the data from tablets and laptops is on Wi-Fi. Wi-Fi carries at least an order of magnitude—perhaps even two—more data than cellular. We typically own only one cellular-connected device but often five or more Wi-Fi connected devices. There are probably around 20 million Wi-Fi access points in a country like the United Kingdom, but only around 60,000 cellular base stations. A hotel or office without Wi-Fi would be seen as unacceptable, whereas one that didn't have cellular coverage would be merely irritating. This is not to underplay cellular; it has a critical role in providing coverage while on the move and will remain an essential part of our communications infrastructure for the foreseeable future.

There are good reasons why Wi-Fi is preferred in most cases. Cellular is expensive to provide and has inherently limited capacity. Wi-Fi is almost free to provide, but we are still some way from reaching the capacity of current systems. This is not because of technology or spectrum—both use nearly identical technologies (such as OFDM) and have nearly identical amounts of spectrum available to them (around 500 MHz in total). The difference comes from the deployment model. Deploying coverage "inside out" is much more efficient than "outside in." With most data usage taking place inside buildings, and with the outer walls of the buildings forming a partial barrier to radio waves, delivering the radio signal from inside the building not only ensures that users have a strong signal but also takes advantage of the isolation provided by the walls to reduce interference to other users. Conversely, cellular systems have to aim to blast through the outside walls, delivering poor signals inside, reducing overall cell capacity, and results in interference to outdoor cells. In principle, cellular could deploy indoors, too—and there have been many attempts to do so using "femtocells" and similar approaches. But the scale of the deployment challenge is beyond a single company and only achievable with Wi-Fi through the actions of millions of users that deploy their own access points. Now that Wi-Fi is widely deployed, the rationale for also deploying cellular indoors is reduced. Hence, a self-fulfilling movement toward Wi-Fi-only devices has happened, since users realize that Wi-Fi connectivity is likely.

This is not an attempt to replace cellular. Wi-Fi can never provide connectivity in rural areas, along most roads, and for most people when they are moving. Cellular is an essential component of our complete communications infrastructure—just not the best way to deliver the final elements needed for ubiquity in most cases.

The remainder of this section considers Wi-Fi changes and additions that might be needed for Wi-Fi to properly fulfill its central role.

Making and Receiving Calls over Wi-Fi

For a device only connected to Wi-Fi (and not via cellular), making a voice call is simple. However, the routing of an incoming call made to the cellular number, which is becoming increasingly commonplace, does require some special treatment.

A simple solution is to use a numbering scheme, not tied directly to cellular, that aims to contact the handset via Wi-Fi—for example, a Skype or WhatsApp "handle" acts in this manner. However, this may not be convenient for the user.

A more complete solution is to make use of enhancements developed by 3GPP (2018), the group responsible for cellular standards. Specification TS23.402—Architecture enhancements for non-3GPP accesses—provides a means for a cell phone to connect to the MNOs network over any Wi-Fi connection by using a secure tunnel. Once connected, the user can both make outbound calls and receive incoming calls over their MNO's network, just as if they were connected via cellular. This does require software support in the handset and the deployment of the relevant security gateway and VoIP support by the MNO. The software has been available on recent releases of the most popular handsets and gateways that are deployed by many MNOs.[3] In the United Kingdom, all MNOs offer some degree of support for Wi-Fi calling.[4]

Automated Passwords

It is normal for travelers to ask "what's the Wi-Fi password?" when checking into hotels, even before they inquire about breakfast and other arrangements. Manually selecting networks and entering passwords is a workable solution but is far from ideal. There are a number of reasons why password protection is used:
- To prevent those nearby from freely using the resource thereby avoiding having to pay for their own broadband connection (or similarly, to ensure that only particular customers benefit from Wi-Fi access).

---

[3] For example, see details of support for voice over Wi-Fi provided by a New Zealand mobile operator (Two Degrees Mobile n.d).
[4] See 4g.co.uk 2018.

- As part of legal restrictions that may require password protection, along with provision of identification, to enable tracking of illegal activities (e.g., the download of copyrighted material).
- To make it harder for hackers to gain entry into the network.

There are also a number of approaches to facilitate entry:
- Common passwords across multiple access points—for example, in all the stores in a chain of coffee shops—or, more broadly, across all access points provided by a particular operator (BT Wi-Fi is a good example).
- Devices that remember passwords and automatically sign into networks on a repeat visit.
- The use of other authentication mechanisms, such as that within a cell phone to authenticate users.
- The use of public access Wi-Fi, such as Hot Spot 2.0.

Logical partitioning of the access point can overcome many of the issues above. Such partitioning—which has no direct connection to the Wi-Fi router owner's network and over which the owner's traffic has priority—can prevent hacking and can make free riders less of an issue.

Modes of operation could be envisioned such that an unknown device is allowed onto a network purely for the purposes of sending an automated registration request, along with suitable credentials. A valid request would receive the password in response, and this would allow full access to the network. A more proactive stance would be for governments or regulators to mandate that all routers sold should make some small fraction of their capacity available for visitors for the limited time that the visitor seeks usage. This will provide access that would be sufficient for most while preventing the problem of long-term "borrowing" of resources.

These measures could create enormous virtual capacity at minimal cost. Such modes and other approaches are likely to steadily evolve over time and can be facilitated by governments by removing regulations that require identification of users for legal purposes.[5] In the interim, users will need to continue to manually enter the password.

---

[5] In most countries there is no requirement for a company providing Wi-Fi connectivity as part of a larger business offering (e.g., a hotel) to record details of those using the Wi-Fi. Security agencies can ask that such a recording be put in place if they suspect the Wi-Fi connection is being used for illegal purposes. Many organizations currently appear to be either misreading the law or applying excessive gating, just in case.

Security

Wi-Fi can provide excellent security as long as appropriate modes of encryption are used. The biggest threat is "rogue" access points. These seem to offer connectivity but yet will inspect data traffic, looking to extract passwords and similar valuable information. Many ways to resolve this could be as follows:
- User applications encrypting data end-to-end to prevent a "man in the middle" being able to extract important information. This is already being done routinely.
- The use of a central validation server. For example, a Wi-Fi device could send the SSID and password used to this server along with other contextual information such as the SSIDs of other visible Wi-Fi nodes. This would allow the validation server to verify that the node is known and has been appropriately certified.
- The use of a system managed by a single company—again, the BT Wi-Fi system is an example.

There does not appear to be any significant security-related reason for Wi-Fi not to adopt a more significant role.

Reliance on Unlicensed Spectrum

Wi-Fi uses unlicensed spectrum that could, in principle, become congested or suffer interference. In practice, we have seen firstly that congestion builds slowly over years, allowing time for it to be addressed, and secondly that regulators have provided additional frequency bands, such as at 5 GHz, when needed. In the future, any emerging problems will likely happen slowly and will be addressed through regulation or something similar.

This does imply that regulators should pay close attention to unlicensed spectrum. With a "Wi-Fi first" policy, spectrum for Wi-Fi becomes more important than that for cellular, and commensurate resources should be devoted to it. This might involve more monitoring to understand congestion and a preference to provide unlicensed spectrum over licensed spectrum. Statements suggesting that the regulator would address issues that reduced the efficiency of Wi-Fi as a matter of great importance would also help reassure users and investors.

More generally, a review of policy toward unlicensed spectrum, and its role and value in the modern environment at both a national and international level, would be appropriate.

Failure of Municipal Wi-Fi

There have been various attempts to cover entire cities with Wi-Fi. These have all failed, mostly because the scale of the challenge is large and the revenues small. The suggestion here is different—not to expand Wi-Fi coverage into areas where there already is cellular coverage, but to selectively deploy Wi-Fi, predominantly indoors, to provide consistency and to be funded mostly by government in various ways.[6]

5G and Wi-Fi

Much of 5G is focused on higher data rates and more increased capacity in dense areas. Earlier chapters have suggested that the delivery of ever higher speeds above the 100 Mbps already theoretically possible with 4G is unnecessary. Delivering increased capacity in dense urban areas would be of value, but the proposed key solution of using small cells and microwave frequencies appears uneconomic and is unlikely to address the majority of data users who are indoors.

Some elements of 5G are useful. The separation of control and data planes and the possibility of better linkage to Wi-Fi could help to form a more seamless use of cellular and Wi-Fi networks. Similarly, the use of software-defined networks (SDN) or NFV could make integration with third-party systems simpler and faster.

Another potential developed that might impact on this area is licensed assisted access (LAA). This is an approach where MNOs use unlicensed spectrum alongside their licensed allocation to improve throughput. Often the control link to the device is retained in the licensed spectrum, with data download occurring opportunistically in unlicensed spectrum. LAA is typically assumed to use the same 5 GHz band as Wi-Fi. LAA is being developed as part of 4G and so might be deployed prior to the 5G era.

Simplistically, LAA might not change much. The handset will choose whether to download its required data from LAA or Wi-Fi. Both use the same frequency bands, and both use similar technology. One would simply substitute for the other with little net effect. However, LAA does allow the MNOs to have more control over the operation of the handset, and this might bring some benefits, especially if some of the developments needed to integrate Wi-Fi and cellular more tightly

---

[6] These might include direct funding, self-provision on government buildings at government's own cost, and indirect funding via license obligations on franchises, or something similar.

fail to materialize. If delivered from a home hub, LAA might provide MNOs with a rationale for a stronger in-home presence; this could also have an impact on the industry dynamics.

Summary

In summary, cellular is already what is used when Wi-Fi is not available—it is our fallback and hence the reason users are not inclined to pay higher monthly fees for cellular connectivity. The approach suggested here to enabling consistency recognizes and builds on this in a pragmatic manner.

Regulatory and Governmental Action

For such a world of consistent communications to happen requires government action of various sorts, as mentioned earlier. Governments and regulators need to change policies away from those focused on speed and toward those aimed at connectivity. Here those policies that are no longer needed are considered, along with the new ones that should be started.

In considering policy and regulatory stance, some thought is needed as to potential industry structure under such a vision. At present, consumers can have a contract for their home line with a company like AT&T and a contract for their mobile with a company like Verizon. Wi-Fi is provided through cable, phone, dish, or other types of companies, or Wi-Fi can be used in satellite hot spots or other hot spots run by companies such as Starbucks. Regulation is typically focused on generating as much competition as possible. In mobile communications, competition is ensured through maintaining three or four MNOs. In broadband to the home or office, there may only be one provider, but competition is generated through unbundled access to the underlying network or other forms of competition above the physical access layer. Success of regulation is measured through access speed and consumer cost with some interest also in universal service provision on fixed lines.

In the future, consumers might also add some form of Wi-Fi access enabler to their set of communication providers. This could be a company like BT Wi-Fi, who actually provides hot spots in some cases, or like Google, who provides passwords and certification of access points deployed by others. They may also have accounts with the government for access in government buildings. The majority of consumer data traffic might flow across this Wi-Fi network. Consumer phone service may be provided by a Wi-Fi access enabler rather than MNOs and be preprogrammed to work effectively using voice over Wi-Fi solutions. Incoming calls might be routed first to the Wi-Fi access enabler and only onto the MNOs if access over Wi-Fi is not available. The contract with the MNO might even be handled by the Wi-Fi access enabler.

Shared network access is likely to grow. MNOs will deliver some of their services across Wi-Fi. High speed connectivity to a railway car's roof to enable Wi-Fi inside the trains might be delivered through a shared network owned by a third party. Similarly, a single rural network that all MNOs use might be constructed.

This is not a radically changed world, but it does have significant changes. Policies need to allow for the innovation and investment needed.

Policies No Longer Needed

Aiming for consistent connectivity would render some current regulatory and governmental approaches unnecessary, including:
- FTTH initiatives and, more generally, a desire to be high in global speed league tables. A universal service obligation set at around 10 Mbps in the home is appropriate, but most home broadband needs can be met via solutions such as FTTC and then VDSL or G.fast over the last drop. More speed requirement takes investment time and money away from areas such as universal Wi-Fi networks that incumbents are typically well placed to deliver.
- 5G test-beds and similar that focused on high data rates would not be needed. Instead, test-beds that improve integration between cellular and Wi-Fi, that demonstrate improved rural connectivity, or that better backhaul to trains would be valuable.
- Competition among the mobile players. Other providers may be more important, and MNOs may be encouraged into network sharing in some cases.

New Policies Needed

The various policies that government needs to embark on have been introduced earlier. These include:
- Investment in Wi-Fi networks in public buildings, including museums, schools, hospitals, universities, and offices in city centers. This includes not only the deployment of the access points but also the introduction or adoption of a universal sign-in system. It should be a relatively inexpensive investment, with access points purchased in bulk and installed by the buildings team.

- Investment in rural cellular coverage through awards of funds against specific coverage objectives.
- Obligations on trains or railway franchise holders and possibly also bus franchise holders to deploy Wi-Fi, with accompanying obligations on track owners to enable effective backhaul provision.
- Potentially greater regulation for Wi-Fi in areas such as spectrum, security, and competitive regulation for any Wi-Fi providers that might have significant market power.
- Potentially, regulation to assist in routing incoming calls to Wi-Fi-connected phones. This could be a modified form of number portability or something similar.

Each of these recommendations are clear and can be embarked on immediately. They typically do not require new legislation, and the funding requirements are relatively modest.

Delivering the Internet of Things

Predictions for the number of connected "things" range from 50 billion to a trillion. There seems little doubt that there is benefit in connecting many of the electronic things around us; this will increasingly happen as the various pieces of the technological and logistical puzzle fall into place.

As discussed earlier, the connectivity to enable the IoT needs not to wait for 5G. Many MNOs have enthusiastically embraced the new technologies emerging within 4G and designed specifically for the IoT, including LTE-M and NB-IoT. Trials were underway in 2016; initial deployment was expected in 2017; and widespread adoption has been expected in 2018. Many IoT implementations can be deployed as software updates to existing base stations, enabling MNOs to quickly and cost-effectively activate IoT solutions across their entire coverage footprint. There are myriad suppliers of chipsets for end-user devices and a strong ecosystem able to deliver software management solutions and other necessary elements.

The IoT may also involve some element of self-provision or of provision by alternative, emerging operators. By analogy, mobile communications are delivered via a mix of operators using licensed spectrum and self-deployment of Wi-Fi access points by myriad entities in unlicensed spectrum. As discussed above, the combination of both is needed to meet the coverage and capacity challenges. Already, a number of unlicensed IoT technologies are being deployed, including those from Sigfox and LoRa and those based on the Weightless standard. Operators such as Arqiva in the United Kingdom and Comcast in the United States are using unlicensed solutions as a way to enter the market, and MNOs such as Orange in France plan to deploy both licensed and unlicensed solutions in a complimentary manner. Some user communities, such as campus-based organizations and airports, are considering deployment across their area. But the path toward a widely deployed unlicensed solution is less clear. It is uncertain which technology or standard will be the winner, or indeed whether there will be multiple winners. It is also unclear whether the route to widespread deployment will be via a single operator or through a community action with shared access. Clarity will eventually emerge, as some solutions gain traction and others fall by the wayside, and bodies such as Weightless[7] aim to assist in the consensus-building process. Others could be more proactive—for example, governments could engage in smart procurement of a nationwide IoT service, spurring suppliers and operators into action. Regardless, the IoT is emerging, and will grow strongly over the next few years.

An Alternative Future

How might such an alternative vision come about? By taking the stance of looking backward from 2022, this chapter now sets out one way that it could transpire. The future is rarely predictable to this degree, and so this prediction is sure to be wrong, but it is useful to demonstrate the possibilities. Here I am assuming that there is less likelihood and longer timescales for Wi-Fi-related initiatives on the basis that no apparent progress has been made from 2016 to 2017. This has essentially resulted in the forecasts being moved out by a year[8] and the emergence of Google's Project Fi on a global basis being discounted.

2018

It was in 2018 that the realization first started to dawn that consumers had gradually moved to a "Wi-Fi first" world. Collecting data on the amount of Wi-Fi usage had always been problematic, but apps running in the background on handsets

---

7 The author, William Webb, is the chief executive officer of the Weightless SIG.
8 Of course, if the forecasts are simply moved out by a year, each year until they finally happen, then these forecasts are not particularly valuable. The need for adjustment illustrates the difficulties of such short time-scale prediction, when events depend on particular strategic decisions made by significant entities such as Google.

showed that over 85 percent of data from a smartphone was sent via Wi-Fi—mostly in the home and office. With over 80 percent of all tablets and laptops being Wi-Fi-only, this implied that well over 95 percent of all of consumer data had been traveling over Wi-Fi, with the percentage growing as Wi-Fi became ever more pervasive and increasingly free.

Wi-Fi projects were enabled around the world. In some countries, such as Singapore, there was already a near-ubiquitous public Wi-Fi network provided by the government. Elsewhere, the Indian regulator TRAI promoted a public open Wi-Fi framework to allow hot-spot owners to offer service to mobile phone users; the European Commission funded and promoted a continent-wide deployment; and myriad local initiatives blossomed.

MNOs mostly ignored the implications. They had much on which to focus, with 5G on the way, promising speeds of over 100 Mbps; with the IoT emerging; with interest in FWA rekindled by 5G smart antenna systems; and with promising partnerships in automotive developing. Further, their strategy teams were kept busy with merger discussions (which were often abortive or blocked by regulators), spectrum auctions, roaming regulations, and more. After all, they were the big players in this space; what did they have to fear from new connectivity models? Early 5G trials, using only a handful of devices, were generally encouraging, helping to persuade all that the 5G vision remained intact.

2019

More gradual change continued in 2019. In some ways, this gradual change was more dangerous than sudden change—the "boiling frog" syndrome occurred, where each little announcement appeared to be of minor significance, and few stood back to look at the impact of all the various activities.

Wi-Fi continued to gain presence, with some governments sponsoring citywide free deployments, and other governments deploying Wi-Fi in all hospitals, museums, schools, universities, and government buildings, making it freely available with only a single sign in needed across all domains on the first time of use. Wi-Fi on trains was also more widely deployed, with some governments making train-operating franchises contingent on Wi-Fi deployment—albeit at a low data rate variant. Wi-Fi gained presence on aircraft and in subways. It was only really when traveling in the car that there was generally no Wi-Fi available—and users often got around this with personal hot-spot solutions.

While Google decided not to deploy Project Fi outside of the United States, other Wi-Fi-oriented MVNOs emerged. During 2019 they found it difficult to access all the features they needed from Android and to assemble the database of

Wi-Fi hot spots, with the result that success was somewhat patchy. In some cases, they were supported by government as a way to further enable access to government Wi-Fi routers and to introduce additional competition.

On the back of increased Wi-Fi availability, the use of voice over Wi-Fi via WhatsApp and similar apps grew fast. This was somewhat aided by MNOs who were finding VoLTE implementation difficult and so were happy to see some voice traffic off-loaded.

The year also saw the publication of a number of user satisfaction surveys that showed an increased understanding that consistent connectivity rather than speed was more important. Users strongly wanted to be connected wherever they were located and were sometimes placed in problematic situations, when using phone-based navigation, for example, and moving out of coverage. Conversely, high speed on cellular was seen as unnecessary, and indeed, after various cases of phone batteries catching fire, manufacturers were also pulling back from powerfully fast processors with high power consumption. Despite this, government officials continued to demand their country be well placed in broadband speed league tables, and MNOs continued to market the 100-plus Mbps data rates that they could offer.

The cellular community was in poor health. MNOs suffering from falling profitability sought various grounds for merger, but regulators continued to block these efforts, convinced that a multiplayer MNO marketplace was the primary route to mobile innovation. The supplier base also suffered—Ericsson and Nokia both announced losses and significant job cuts, and even Huawei experienced a fall in mobile network infrastructure sales. The news from 5G trials became less positive as more widespread deployments resulted in interference cases. This led to a growing realization to the real-world deployment complexities and waning interest from verticals in the service.

Cisco's VNI market projections, published midyear, were the first ever to predict that data growth on cellular would be minimal over the coming years. This was based on the observation that data volumes had stopped growing in many markets, for reasons that included attempts by MNOs to increase prices, increased off-load to Wi-Fi, and a degree of saturation of uses for a mobile phone. Data growth over Wi-Fi, however, was still predicted to be about 30 percent to 40 percent per year.

In the world of IoT, the cellular community failed to gain much traction. NB-IoT and LTE-M were slower to appear than expected, and unlicensed technologies finally grouped together into a standards body to present a unified front with a common, single chipset that was embraced by a wide range of device manufacturers.

2020

With hindsight, it was clear that 2020 was a pivotal year where the cellular community realized its vision of the future was at serious risk, and key players broke ranks. Across many countries, MNOs that had significant Wi-Fi presence in the United Kingdom (such as BT, which owned EE) offered a Wi-Fi first package where their devices worked seamlessly across Wi-Fi and cellular, transferring data wherever possible via Wi-Fi. By doing this, such companies were able to offer lower cost unlimited data packages at a time when other MNOs were increasing the prices of their premium bundles.

The 5G vision became ever more confused as various countries and operators deployed their own variant of a semi-proprietary solution for services such as FWA. Meanwhile, regulators failed to agree on consistent frequency bands. Many claimed their 4G solutions were, in fact, 5G, while the visions of mmWave services and low-latency solutions were quietly deprioritized. Where super-fast mobile services were offered for increased cost, the take up was minimal, leading MNOs to offer them at no premium. Most subscribers were uninterested, having seen no benefits from previous speed upgrades.

In addition to the MNOs, others now entered the market to provide connectivity solutions across multiple platforms and operators. Amazon launched their offering worldwide as did Facebook and WhatsApp.

With Wi-Fi becoming ever more pervasive, regulators clarified and changed unhelpful laws requiring Wi-Fi sign in from business providers. This meant that automated sign in was now widespread—the days of asking for a Wi-Fi password or agreeing to terms and conditions of access were over. Some regulators also started to monitor congestion in Wi-Fi spectrum and pledged action should Wi-Fi frequencies become congested—effectively giving Wi-Fi a quasi-licensed status. In practice, with ever better frequency management solutions, Wi-Fi congestion remained minimal in all but a few major train stations and stadiums.

There was increasing outcry among MNOs that regulators were heavily constraining them but had no equivalent regulation for the connectivity providers, resulting in a situation where the MNOs were unable to react effectively. Privately, regulators acknowledged this, but the pace of regulatory change was slow, and few were willing to admit their focus on competition was no longer working.

2021

In the first quarter of 2021, more subscribers signed up to connectivity platform providers than to MNOs for the first time. With declining subscriber numbers, a few major MNOs started to withdraw from the market, selling to their competitors, with regulators finally allowing this to occur when they realized there was no other option. Other MNOs started to move toward a wholesale-only model, shutting down shops and reducing customer services. They also withdrew from the IoT marketplace, unable to cost-effectively serve most of the customers, unwilling to make long-term commitments on service in very uncertain times, and undercut by unlicensed providers using simpler technologies.

There was little interest in widespread 5G auctions for mmWave spectrum in the 24 GHz and 28 GHz bands, with only reserve prices being met in many countries. All work on mmWave systems was quietly shelved, even in academia.

Public access from individuals and households who shared their Wi-Fi using a second identity that was protected by a firewall became the norm.

New regulatory guidance from the European Commission and FCC recognized the changed world and suggested that regulators focus on connectivity platform providers rather than MNOs across areas such as spectrum, competition, consumer protection, and more. There was increasing concern that the connectivity providers were able to negotiate very favorable deals with wholesale MNOs, leaving the MNOs little revenue to invest in network expansion.

2022

By 2022 the cellular industry was very much the underlying bit pipe—the bottom of the value chain. MNO brands started to disappear from main streets and no longer took out advertising. Few offered direct subscriber contracts anymore. Most countries consolidated down to two MNOs, often with some sharing of assets between them. Instead, consumers looked to Google, Apple, Amazon, and others for their connectivity contracts. These companies provided connectivity at lower cost with unlimited data, free roaming, and better coverage.

Substantial changes occurred to research and regulatory activity. It now concentrated on Wi-Fi and how to ensure seamless links across to cellular where Wi-Fi was not available. Bodies such as 3GPP and the GSMA refocused, and even the Mobile World Congress was rebranded as the Global Connectivity Congress. It was here where keynote speeches from chief executive officers of MNOs and equipment manufacturers reflected on how it could all have been so different.

Conclusions

This chapter has discussed how speed of data connection is now becoming less important than consistency—the ability to be connected at a reasonable speed everywhere. Rather than aiming for ever faster connections, it suggests that delivering enhanced coverage in a number of known problematic locations such as trains and rural areas would generate greater value for the economy and be preferred by most consumers.

In most of these locations Wi-Fi is a better solution than cellular, with the exception of coverage in rural areas. This reflects a trend that has been underway for years toward increasing use and reliance on Wi-Fi to the extent that it is now the preferred method of communication. Developing policies for a "Wi-Fi first" world is becoming increasingly important for governments and regulators.

The chapter has also discussed how the IoT community is well on its way to delivering the connectivity required and that little intervention is needed—although some targeted procurement might speed things along.

The end result—connectivity everywhere—would be a goal well worth striving for. A great road infrastructure is no longer one with unlimited maximum speed, but one with minimal congestion and excellent safety. A great communications system is one available everywhere, at all times, with minimal congestion and at low cost. If the focus of 5G could be switched to this direction, it would be a new generation worth having.

Chapter 8 References

3GPP. 2018. Specification TS 23.402: Architecture Enhancements for Non-3GPP Accesses. Release 15, Version 15.3.0, March 27. https://portal.3gpp.org/desktopmodules/Specifications/SpecificationDetails.aspx?specificationId=850.

4g.co.uk. 2018. "Which UK Networks Offer Wi-Fi Calling?" *4g.co.uk.com*, January 23. https://www.4g.co.uk/news/ee-o2-three-and-vodafone-which-networks-offer-wi-fi-calling/.

Australian Government, Department of Communications and the Arts. N.d. Mobile Black Spot Program. Website. https://www.communications.gov.au/what-we-do/phone/mobile-services-and-coverage/mobile-black-spot-program.

European Parliament. Procedure 2016/0287/COD. COM (2016) 589: Proposal for a Regulation of the European Parliament and of the Council amending Regulations (EU) No 1316/2013 and (EU) No 283/2014 as regards the promotion of Internet connectivity in local communities. TFEU/art 172. https://eur-lex.europa.eu/procedure/EN/2016_287.

Two Degrees Mobile. N.d. "Wherever there's WiFi, there's 2degrees." Website. https://www.2degreesmobile.co.nz/help-and-support/mobile/products-and-services/wifi-calling/.

Chapter 9
Regulation, Competition, and Broadband

There are wider factors shaping the future communications environment than just mobile manufacturers and MNOs. The environment within which they exist is increasingly shaped by regulatory policy—controlling mergers, obligations, prices, and competition to varying degrees.

As has been seen in Chapter 8, there are alternatives to classic cellular communications in the form of Wi-Fi connected to home and commercial broadband. Fixed connections are a core part of our communications environment. It has been said that every connection is a mix of wired and wireless, and the only variation is where the wires stop. So for cellular, there is typically a wired connection from the core network to the base station and then wireless after that. Wi-Fi is wired to the router and wireless for the last few meters. As cells get smaller, the wired element gets larger and the wireless, smaller.

This chapter looks at the wider environment and the impact it might have on how 5G transpires.

Regulation and Competition

The world of cellular communications is highly regulated. Access to a key input—spectrum—is controlled by spectrum regulators around the world such as the FCC and Ofcom. They often place conditions on license awards to bring about outcomes such as increased competition or improved coverage. As mobile communications have become part of the critical national infrastructure, regulators and governments have placed increasing pressure on MNOs to deliver reliability. Equally, governments are keen to see new services roll out in their country and encourage MNOs to deploy the latest generation of mobile technology via early license awards, exhortations, and, in some cases, specific policies.

For many regulators, the most important factor is the maintenance of competition. For example, Sharon White, the CEO of Ofcom, has said:

> Competition is the lifeblood of today's telecoms market, spurring innovation, better coverage and fair prices. Just as President Hoover observed: "Competition is not only the basis of protection to the consumer, but is the incentive to progress." (White 2016)

Regulators believe that competition is a fundamental good in the industry as it results in innovation and lower consumer prices and enables regulators to take a light-touch approach to controlling the sector. (Conversely, for fixed-line communications, where there is rarely effective competition, regulators often take a very interventionist approach—setting prices, controlling company structure, and having an input on investment strategy.)

This is not the right place to discuss the debate as to whether this focus on competition is appropriate. Here I am merely noting its impacts:

- *Mergers are generally prevented.* Any mergers that reduce the number of MNOs would appear likely to lessen competitive pressures (although there is evidence to the contrary from Austria and elsewhere). For that reason, in most countries, attempts by MNOs to merge are declined by the relevant competition authorities. For example, in 2016 the MNO Three attempted to merge with O2 (owned by Telefonica) in the United Kingdom. The merger was referred to the European Commission. In the meantime, the UK regulator Ofcom and the UK's Competition and Merger Authorities both issued statements showing they did not support the merger. The European Commission eventually decided to block the request on grounds of reduced competition. Those in the industry have the view that mergers within a country will be impossible for the foreseeable future.
- *Fully shared networks are discouraged (although partial sharing is encouraged).* Broadly, regulators would prefer that MNOs had their own networks. This maximizes competition. However, it has been recognized that the economics of four or more nationwide networks is increasingly difficult to support, and most have allowed, or even encouraged, a degree of network sharing where MNOs can share masts and associated elements such as power and backhaul. Most regulators draw the line at sharing spectrum, insisting that each MNO only transmits in its own spectrum and not provide transmission services for others. Regulators fear that if full sharing were allowed, it would open the door to a single network operated by a third party on behalf of all the MNOs, with the resulting loss of competition. However, as discussed earlier, there are elements of 5G—such as the proposal for dense mmWave deployments—that would only appear viable if full sharing were possible. Rural coverage would also benefit from full sharing (or nationwide roaming). This is a position that may be more amendable to change as the economics of 5G become clearer.
- *Innovation in business models is constrained.* While regulators often target innovation in technology, trying to ensure that they pave the way to the next generation, they often inadvertently block innovation in business models through their concern about reduced competition or about the impact on the consumer. This can happen via outdated regulations, such as requirements

to maintain emergency call capability, or via other concerns such as net neutrality. When such issues become apparent, regulators will often address them; but the regulatory process is slow, involving debate, consultation, and statement. It can take many years to address issues, and by that time nimbler competitors may have delivered an alternative.
- *Sector profitability is reduced.* Competition directly reduces profitability. In addition, regulators have placed further pressure on revenues with initiatives such as restrictions on roaming tariffs (in the European Union), coverage obligations, and other regulatory burdens. When the industry was highly profitable, such interventions could be borne by the MNOs. But with profitability reduced, as set out earlier, further milking of the sector reduces the scope for investment. There is always a trade-off between delivering the lowest prices for consumers and enabling risky investment in future networks. If there is a need for the MNOs to make significant infrastructure investments, then it may be that the balance has shifted too far toward short-term consumer benefits.

Regulators tend to spend a large part of their effort focused on the mobile sector. This is unsurprising, given the size and importance of the sector, but it may be inappropriate in a Wi-Fi first world (as introduced in Chapter 8). For example, large teams are concentrating on clearance and auction of spectrum for the next generation of mobile, whereas the spectrum used by Wi-Fi gets much less focus. Such a focus should benefit 5G at the expense of other solutions by ensuring that it gains the spectrum it needs. However, as mentioned earlier, there is perhaps too much attention here. This has resulted in competitive positioning between regulators keen to be the first to provide 5G spectrum in their country to deliver a commercial advantage to their manufacturers and operators. This is most apparent in the FCC's 2016 move to open up 28 GHz for mmWave access.

Chapter 4 set out the many difficulties with mmWave solutions, including the need to focus industry effort on cost reduction and innovation in radio components working at mmWave frequencies. This would have been best achieved with global agreement on a preferred frequency band. Such agreement often takes time and diplomacy, as different countries have differing legacy uses and time and studies are needed to understand the impact of changing allocations. It is not unusual for a new band to take five or more years to reach global agreement. Debate had started on the optimal mmWave band, but before consensus could be reached, the United States decided to unilaterally move ahead with 28 GHz—a band that suited their current use of spectrum and has also found some favor in early research activities. However, this band is highly problematic elsewhere, as it is used for satellite communications where any change of frequency is impossible once the satellite is launched. It remains to be seen whether other countries will eventually feel that they have to follow the US's lead, or whether there will be an alternative band suggested in other parts of the world. Regardless, this fragmentation is deeply unhelpful for 5G mmWave evolution.

There is little to suggest to regulators that they should change their current position. The impression painted by the mobile industry is that 5G is in robust good health and that MNOs are keen to deploy it as soon as possible. Although the picture is very confused, most expect 5G to be delivered in just the same way as previous generations—via MNOs focused on upgrading their networks encouraged by competitive pressure. This is one of the downsides of the "emperor's new clothes" situation outlined in Chapter 6. Had the industry painted a picture of a more difficult introduction of 5G, it would have paved the way for discussions with regulators on how to ease competition and overcome issues such as planning permission. *Talking up 5G has made it harder to introduce 5G.*

In summary, the industry is likely to have an inflexible structure as it enters the 2020s and the period when 5G might be deployed. Without a reduction in competition, MNO finances will remain weak, and their ability to use novel approaches such as shared networks will remain limited.

Radio Spectrum

All new generations of cellular technology have had new spectrum associated with them. For 2G it was frequency bands at 900 MHz; for 3G, at 2.1 GHz; and for 4G, at 800 MHz.

Some have noted that 3G was an unsuccessful generation, in that MNOs probably failed to make returns on their 3G investments. This may be in part because of the high frequency associated with 3G. It required many more base stations which came with an associated cost. Conversely, 4G returned to a lower frequency band, enabling the economics to work better.

For 5G the frequency bands to be used are far less clear, but insomuch as there is a trend it is toward 3.4–4.2 GHz and mmWave bands at 24–30 GHz.[1] These are clearly all well above the frequency of 2.1 GHz that caused problems for 3G

---

[1] In addition, in some countries the 700 MHz band is assumed to be for 5G. However, this has already been auctioned in the United States, Germany, France, Australia, and elsewhere. The total bandwidth of around 100 MHz means that, if divided among say four MNOs, they would each have two times 10 MHz (with the remainder being used for guard bands). Many already have similar amounts, both at 800 MHz and 900 MHz, so it is hard to see how this could make any material difference.

and would result in much reduced coverage of 5G compared to existing generations. However, the linking of generations to frequency bands is now less strong. In the bands previously used for 2G and 3G, 4G is now being deployed, and it may be that 5G is likewise used in frequencies already owned by the MNOs. But many MNOs are in the process of refarming their 2G and 3G technologies to 4G by deploying 4G in these bands. They would be disinclined to rapidly refarm these to 5G, preferring to leave 4G in the bands for many years, perhaps even a decade, to gain a good return on their investment.

As a result, the spectrum position is not a good one for 5G. There is a lack of consensus globally on 5G spectrum allocations, leading to fragmented economies of scale and slow introduction of equipment and devices. The spectrum identified is at relatively high frequencies, making extensive coverage unlikely. Refarming of existing holdings is possible but MNOs will be disinclined to do this, and it may take many years before 5G equipment is available across the 40 or more frequency bands used around the world by the MNOs.

The cellular spectrum position is far from ideal for 5G—the spectrum position will slow deployment and tend to restrict it to urban areas.

Regional Differences

The approach to 5G is not homogeneous around the world. While the same technology would be adopted, and the same services likely to be used, governments and regulators often have very different approaches. Also, larger countries such as the US and China have sufficient economies of scale that they can pursue national frequency bands and slight local variations. For example, the differences in approach between the US and Europe are shown in Table 9.1.

Table 9.1: Differences between the United States and Europe

United States (FCC Chairman, June 2016)	Europe (European Telecom CEOs, July 2016)
Now, I've listed some examples of what 5G makes possible, but if anyone tells you they know the details of what 5G will deliver, walk the other way.	The European 5G Action Plan must reassure vertical industries that 5G deployment will be synchronized across Europe to achieve homogeneous availability both in terms of location and time.
We will be repeating the proven formula that made the United States the world leader in 4G. It's a simple formula: Lead the world in spectrum availability, encourage and protect innovation-driving competition, and stay out of the way of technological development.	The European Commission and Member States must encourage and incentivize cross-sector innovation through cross-sector hubs for experiments, trials and large-scale pilot projects.
Unlike some countries, we do not believe we should spend the next couple of years studying what 5G should be, how it should operate, and how to allocate spectrum, based on those assumptions. Like the examples I gave earlier, the future has a way of inventing itself. Turning innovators loose is far preferable to expecting committees and regulators to define the future. We won't wait for the standards to be first developed in the sometimes arduous standards-setting process or in a government-led activity.	European stakeholders [will] agree on trial specifications valid for pan-European trials [and] will target launching 5G in at least one city in each of the 28 European Member States by 2020. The European Commission should establish a 5G Venture Fund to foster a new wave of start-ups and innovation around 5G technologies.
I would also emphasize that the development of 5G is not anything like an international zero-sum game. Rather, it is a contest in which everyone can win. Our success and that of others redounds to the benefit—literally—of everyone in the world.	The European Commission intends to develop a 5G Action Plan aimed at giving Europe the leadership in the deployment of standardized 5G networks.

Sources: United States: Wheeler 2016; Europe: BT Group et al. 2016.
Note: I give thanks to Richard Feasey for assistance with this table.

Clearly, the United States is biased toward the support of innovation, while Europe is more focused on harmonization and the establishment of test-beds and such. Historically, the US approach has been more successful in fostering companies like Qualcomm and, in recent years, in the relatively early deployment of new technology. However, it is unclear whether this will hold true in a more uncertain future.

Some Asian-Pacific countries such as China and Korea tend to have a more interventionist approach, where governments seek to ensure advantages for their local manufacturers. This can be seen in the desire for early deployments in South Korea and in various industrial plans in China, Korea, and Japan. MNOs in these countries are more likely to deploy some 5G solutions early in order to meet government expectations, even if they do not expect it to be profitable.

None of these differences change the underlying problems with 5G discussed in earlier chapters. However, they do change the likelihood of MNOs deploying some 5G elements at an early stage—for example, there might be early mmWave test-beds deployed in the Asian-Pacific region.

Broadband Fixed Access

With the advent of almost every new wireless technology, someone suggests that it can, finally, realize the vision of FWA—the idea of using wireless to provide a broadband pipe to the home rather than copper or fiber. It is no surprise that the forthcoming arrival of 5G, whatever and whenever that might be, has prompted some to suggest it is the solution to FWA—in particular, Verizon has pushed hard in this direction, and many manufacturers such as Nokia see it as a key 5G usage case. Will 5G finally crack the FWA conundrum?

It is worth first reminding ourselves of some history. FWA really came to the forefront around 1996 as GSM acquired data rates as fast as the then-best fixed line speeds. High profile launches included Ionica in the United Kingdom, who exploited a purpose-built technology from Nortel. But by 2000, all had failed. The costs of FWA deployment in the real world proved much higher than expected, and the telecommunications providers reacted by upgrading their fixed lines and reducing their prices. Since then there have been myriad new attempts, such as Clearwire (2.5 GHz WiMAX [worldwide interoperability for microwave access]), Verizon HomeFusion (LTE), PCCW UK Broadband (3.5 GHz WiMAX), and others. All failed to gain any significant number of subscribers. There have also been attempts to use frequency bands above 20 GHz—the UK's Radiant pioneered a new technology with mechanically steerable antennas that could form a mesh, but deployment proved harder than anticipated. Motorola had a solution based in the local multipoint distribution service (LMDS) bands, but again that proved too expensive.

Has anything changed since then? Data rate expectations to the home have steadily increased, as have data volumes, providing a moving target that wireless is struggling to keep up with. Wired home broadband services that deliver in excess of 50 Mbps and provide more than 50 GB per month are now commonplace in developed countries. And with new solutions such as G.fast on the horizon, data rates are likely to continue to grow for the coming decade. Prices have stayed approximately constant despite the ever improving service, making the economics more challenging.

Does 5G bring anything new to the table? As discussed in Chapter 7, it is not really known what 5G will entail, so this question is hard to answer. But it may have a mmWave component with beam-forming antennas and the capability to deliver data rates in excess of 100 Mbps. This, in principle, makes it competitive with wired connections on data rates. But mmWave systems have very limited range—typically 100 meters or so; although, with directional antennas on both ends of the link and line-of-sight propagation, 500 meters might be viable, as long as it is not raining heavily. The short range was the reason Radiant went for a mesh solution, with connectivity bouncing from house to house. If 5G is widely deployed, then economies of scale might also reduce equipment cost. Beyond that, there is little that is new.

The problem with short range is that it means the system is best suited for urban and suburban areas. In rural environments the house density is too low for there to be more than one or two homes per base station, making the system uneconomical. But it is in rural areas where the problems of connectivity are most acute and the demand for alternatives highest. Rural areas are often best served by wireless solutions in the lowest frequency bands because they have a long range—one of the reasons TV white space (see Webb 2013) was seen as a possible game-changer for FWA in rural areas (and solutions of this sort are being deployed in the United States). In urban and suburban areas wired connectivity is often already good, so the wireless solution has to compete on price or on data rate, or both. Urban areas are often so cluttered that line of sight is problematic for many homes within the nominal coverage radius of a base station. Base station locations are hard to come by because the best have already been taken by mobile operators. Suburbia might be a suitable compromise location, but the competition issues still exist.

The economics of FWA are harsh. A base station might cost around $20,000—more if the antenna array is expensive—and another $20,000 to deploy. Site rental and backhaul costs can easily be $10,000 per year. Home installation is typically another $300 or so for the equipment and $200 for the installer to fit and align roof-top antennas. On top of that is marketing, which needs to be substantive to persuade home owners to switch providers, perhaps adding $200–$400 to customer acquisition costs. US suburban density is around 1,000 people/km^2, or perhaps 400 homes/km2. With a coverage range of 500 meters, a base station covers around 75 percent—or 300 homes. If 10 percent could be persuaded to switch and were able to be connected, then there would be 30 homes per base station. Total costs per home would be around $2,000 initial expenditure and $350 per year operational costs. Amortized over 10 years, that is around $500 per home per year. A $50 per month service just breaks even. Of course, penetration rates higher than 10 percent might be achieved, but with 30 homes all potentially requiring 100 Mbps during busy-hour periods[2] in the next few years, the resulting 3 Gbps capacity requirements will stretch even a 5G base station.

Interestingly, outside of the United States, broadband packages are around $40 per month but in the United States prices can reach $100 per month. This suggests that either there is an opening in the United States for FWA that does not exist elsewhere or that competitive forces have yet to reduce US broadband costs to those in the rest of the world. Verizon appears to be betting on the former, in its desire to roll out a mmWave FWA solution based on its own specifications, that it is claiming is 5G. Perhaps there is something special about the United States that will allow FWA to find a role there. Or perhaps their efforts will be added to the long list of failed FWA initiatives.

Verizon's initiative is destabilizing to the wider 5G effort. Verizon is planning to use 37 GHz—yet another different frequency band—for their deployment. They have written their own specification that they have claimed is 5G, and this will also add confusion and fragmentation.

Conclusions

This chapter has shown that regulatory forces and governmental policy also have an impact on the mobile sector and hence on the form and timing of 5G's introduction. While regulators profess a strong desire to promote innovation and new technologies, in practice their focus on competition is likely to undermine the ability of MNOs to find innovative solutions to the problem of financing 5G deployments. A better regulatory approach would be to allow mergers, the deployment of shared networks, and the emergence of various OTT and MNVO-like models. But given the impression portrayed by the industry that 5G is thriving and imminent, it is unsurprising that regulators see no need to change their current positions—indeed, they might conclude that these positions are helping

---

[2] This would imply that all users would all want large bandwidths simultaneously. In most networks this rarely occurs. However, in homes where video is primarily consumed via broadband connection, it could be imagined that most households would have some video streaming during the evening; hence loading across a cell may have a higher correlation.

to speed 5G implementation. The net effect will be unhelpful, but this will only become apparent over the next few years.

The chapter has also considered whether 5G might have a role to play in broadband FWA. Some US companies believe that it does, and the particular economics of the United States may favor it there. However, the history of FWA is bleak, and there is little in 5G that would seem to change the underlying dynamics.

This concludes the exposition of the 5G myth. Chapter 10 summarizes the arguments set out in this book.

Chapter 9 References

BT Group, Deutsche Telekom, Ericsson, et al. 2016. *5G Manifesto for Timely Deployment of 5G in Europe*. Brussels, Belgium, July 7. Available at http://telecoms.com/wp-content/blogs.dir/1/files/2016/07/5GManifestofortimelydeploymentof5GinEurope.pdf.

Webb, William. 2013. *Dynamic White Space Spectrum Access*. Cambridge, UK: Webb Search Ltd., September. http://www.webbsearch.co.uk/wp-content/uploads/2013/09/Dynamic-White-Space-Spectrum-Access-by-William-Webb.pdf.

Wheeler, Tom. 2016. "The Future of Wireless: A Vision for U.S. Leadership in a 5G World." Prepared remarks of FCC Chairman Tom Wheeler, National Press Club, Washington, DC, June 20.

White, Sharon. 2016. "Ofcom Comment on the Proposed Merger of Three and O2." *Ofcom*, February 1. https://www.ofcom.org.uk/about-ofcom/latest/media/speeches/2016/three-and-o2-merger.

Chapter 10
How the Future Plays Out

Why 5G As Currently Envisioned Is Flawed

The key underlying rationale for previous generations and for 5G has been to meet ever growing user requirements for more data and faster connectivity. Chapter 3 suggests that this trend is coming to an end. Current mobile data speeds are more than adequate for all foreseeable uses. Data growth is slowing and may plateau around 2027, with only about two times the growth occurring during the 5G era. With 5G predicated predominantly on higher speeds and also on its ability to deliver substantially enhanced data capabilities, this suggests that 5G may not be targeting the right areas.

Chapter 4 shows how technology has improved dramatically over previous generations but further improvements are hard-won. This broadly means network enhancements become expensive in the form of many more antennas at the base station and in the device, many more small cells, or dense deployments in completely new frequency bands. All of these are uncertain; some are untried, and some will require substantial further development. The advent of 4G effectively provided a capacity enhancement of around 2.5 times at very little extra cost. The same will not happen for 5G. Relatively low-cost capacity enhancements will likely provide less than two times the improvement. Going beyond this will come at a very high cost due to the very large number of additional small cells that will need to be deployed.

Chapter 5 shows that the economics of the mobile industry have changed substantially over the decades. From a time during 2G when the MNOs were some of the most profitably listed companies, they have fallen to the point where they are underperforming the "all sector" benchmarks by some 50 percent. Revenues are not expected to rise, while investment is anticipated to continue at relatively high levels. The only rationale for MNOs to invest in new technology is to prevent subscriber churn to their competitors. This threat has resulted in their moving quickly to deploy 4G, which does have material benefits for subscribers. But without any clear benefits from 5G there is limited incentive for MNOs to upgrade their network.

Chapter 6 demonstrates that it is in the interests of all the key players to be supportive or even strong promoters of 5G. Academics benefit from 5G initiatives as sources of funding. Manufacturers rely on the rollout of 5G to provide a boost in revenues. Operators fear if they step out of line they will suffer competitive disadvantage. Governments see political benefit in being supportive. It is in nobody's interest to rock the boat.

The 5G community cannot be accused of being short of visions—quite the converse. Chapter 7 discusses visions which range from metrics for the radio system to a wide breadth of use cases. There's the feeling that 5G is intended to solve all the problems of the mobile community and provide a utopian solution where all have perfect communications that meet every need that users might have. This compares with previous generations, where the visions have been much more restricted—such as improving capacity or providing a specific data rate.

But these visions are too utopian. Fully achieving them would not only require astonishing breakthroughs in radio technology but also require subscribers to significantly increase their spending. Both are heroic assumptions. In practice, most visions can be adequately achieved with existing technology such as evolved 4G, evolving Wi-Fi, and emerging IoT technologies.

The future is uncertain, and requirements or services may well appear that result in a different industry than seen today. But until that happens, 5G investment cannot be justified.

A Better Vision: Consistent Connectivity

Chapter 8 discusses how speed of data connection is now becoming less important than consistency—the ability to be connected at a reasonable speed everywhere. Rather than aiming for ever faster connections, it suggests that delivering enhanced coverage in a number of known problematic locations (such as trains and rural areas) would generate greater value for the economy and be preferable to most consumers.

In most of these locations Wi-Fi is a better solution than cellular—with the exception of coverage in rural areas. This reflects a trend that has been underway for years toward increasing use and reliance on Wi-Fi, to the extent that it is now the preferred method of communication for most people. Developing policies for a "Wi-Fi first" world is becoming increasingly important for governments and regulators.

The goal of policies such as this is to deliver consistent connectivity—the ability to be connected everywhere with a quality sufficient for the majority of applications we use on the phone. This would deliver huge value to consumers, be in line with what users of mobile phones see as most important, and deliver a new generation of phones that would be materially better.

Chapter 9 considered regulation and its role on the mobile sector and concluded that regulation tended to stifle innovation. It also noted that the current focus on competition might be inappropriate for a future where less competition but more investment would be more likely to lead to a desirable outcome. However, the chapter was pessimistic on the ability of regulators to see this and to change their policies and beliefs in the near-term, further impacting the likelihood of a successful 5G deployment.

The chapter also considers whether 5G might have a role to play in broadband FWA. Some US companies believe that it does, and the US economy may favor it there; but the history of FWA is grim. There is not that much change in 5G that would affect the underlying dynamics.

Significant Industry Structural Change

The analysis of 5G has shown that the mobile industry is in relatively poor shape. Revenues are static and profitability is poor relative to other sectors. Regulation is backward looking and not favorable toward innovation or structural change. Coverage is not improving materially and some areas, such as trains, continue to have poor coverage over 30 years after the introduction of 2G. Other forces such as the growth of Wi-Fi are emerging.

It could be that 5G becomes a catalyst for industry change. Indeed, it might be that there is no new 5G technology, but the inability to introduce it allows for debate and change in the industry. In debates, over 80 percent of delegates expected that the 2020s would see very significant structural change in the industry, although there was less clarity about what these might be (Cambridge Wireless 2016a,b). Many realize that something has to give.

Earlier chapters have hinted at what this structural change might look like. These include:
- The boundary between MNOs and OTT providers shifting, with OTT providers taking on more of the functions of the core network.
- The radio access network (RAN) consolidating to one or two providers, with single shared networks in hard-to-cover areas.
- Wi-Fi playing a wider role, with simplified access to millions of access points per country.
- Alternative communications providers emerging and offering service across the RAN networks and Wi-Fi, coupled with other services such as the Google suite of mobile products.

This would lead to a dramatic shift in power within the industry:
- Manufacturers of cellular network equipment would suffer as the number of RANs declined; there would be no new technology. Some will cease to exist in their current form.
- Conversely, manufacturers such as Cisco will benefit, as the core elements become more important and are duplicated across multiple parties.
- MNOs will suffer. Many will merge as the number of RANs drops. They will have falling consumer presence.
- OTT providers and new entrants will benefit by gaining subscribers.
- The Wi-Fi community will benefit as Wi-Fi becomes an ever more important part of the national infrastructure.

This would be a seismic shift, and one that has the potential to restore the industry to profitability. It would facilitate a raft of innovation, changing our communications landscape beyond recognition.

It may transpire that 5G is not a new technology or set of services but a catalyst for dramatic industry change after years of gradual decline.

5G Becomes Whatever New Stuff Happens

After all of this discussion, is it any clearer what 5G is?

This book has shown that 5G cannot be what is currently claimed. The visions of 10 to 100 times faster speeds and 1,000-time increases in capacity as put forth by the key players are unrealizable, and the technology to deliver a new generation has not materialized. As currently proposed, 5G is a myth.

One possible outcome is that it takes the industry many years to realize this. MNOs deploy some 5G elements, such as mmWave city center systems and ultralow latency solutions, only to discover that there are few services that require them. Investment slows and promotion of the new capabilities disappears. The 5G deployments would be increasingly mothballed and the industry would come to realize that the experience was an expensive mistake. This outcome is most likely in the Asian-Pacific area where MNOs have stronger balance sheets and where there is a powerful culture of being at the forefront of new technology. This can be seen in the desire of Korea Telecom to have a 5G solution at the 2018 Winter Olympics and the push from operators in Japan and elsewhere to lead in 5G. It is less likely in Europe where MNOs are under severe investment constraints, and unlikely in the United States where MNOs are more pragmatic and less visionary.

A much more likely outcome is that there is little real 5G deployment, but the industry saves face by claiming that 4G systems that are currently being implemented are really 5G. There are already signs, for example, that some manufacturers are proposing that the implementation of NB-IoT coupled with a virtualized core would comprise a 5G solution. These developments are very much 4G solutions implemented nowadays on current 4G networks. But 5G is ultimately just a name. Anyone can and will claim that they have a 5G network, even if they have just implemented the latest 4G upgrade.

During 2017 the view increasingly emerged that 5G would initially be deployed as a "non-stand-alone" radio within a 4G solution. Here, operators deploy a 4G system in bands such as 3.4–3.8 GHz on the most congested 1 percent to 2 percent of their cells but reserve one of the carriers for a 5G interface. The 4G network continues to provide all of the signaling and control but can direct suitably equipped mobiles onto the 5G carrier for their data. This might initially require a software update to the network, although the resulting benefits will be minimal—perhaps a slightly reduced latency and data rates extending above 100 Mbps in city centers. In fact, most users would not notice any difference for this sort of 5G connectivity—a far cry from the vision that 5G would revolutionize communications.

Some have suggested that 5G will be whatever interesting developments happen from 2018 onward. This appears highly plausible.

Calling whatever is deployed in circa 2018 a "5G network" makes political sense, even if it is nonsense from any logical viewpoint. There has been much political capital expended in claiming that 5G will be deployed early in a country. MNOs and governments will simply claim from 2018 onward that they have 5G, even though all that has been deployed is evolved 4G. For all the debate, 5G could just be a label—not a technology. End users will be told that they now have 5G even though they have not changed their handsets or received any improvements in service. Alternatively, in the case where 5G is the label applied to the introduction of IoT, consumers may be told that 5G is not about their handset but about the ability to connect their devices. In the United States, the term 5G might be applied to FWA deployments, with "5G to the home" competing with FTTH. Given the confusion around what 5G is, this second "label whatever we have as 5G" approach could be used in Europe and the United States alongside the first "limited mmWave deployment" approach happening in Asia-Pacific.

But simply labeling current developments as 5G is an opportunity lost. As discussed in Chapter 8, there is much change that both benefits consumers and leads to a more sustainable structure for the industry. If governments, regulators, and MNOs stop competing on having the fastest solutions in the world and refocus on consistent connectivity, then 5G could be a strong force for good.

The Future Is Bright—Once the Vision Is Realigned

This book has been predominantly critical of the current direction of the mobile communications industry and, in particular, the vision of 5G as portrayed by key players. But this book has not been negative about the potential, future benefits of mobile communications. Delivery of IoT nationwide would be transformational and one of the greatest achievements of the industry for decades. Delivery of consistent connectivity would be appreciated by almost all mobile phone and internet users. Collectively, it would be more than a new generation—effectively a new era in wireless—and it can be readily achieved with appropriate focus. In doing so, many of the problems that ail the industry can be addressed. All we need is a little realignment. The aim of this book is to stimulate debate that allows 5G to become a good thing that happens from 2018 onward. After all, it's good to talk, and the future can be bright.

Chapter 10 References

Cambridge Wireless. 2016a. "Debate 1: Technology Readiness." The CW 5G Debate in association with the NIC held at the Shard, London, United Kingdom, October 24. https://www.cambridgewireless.co.uk/media/uploads/resources/Debates/24.10.16/5G_Debate1_24.10.16_FinalTranscript.pdf.

Cambridge Wireless. 2016b. "Debate 2: Business Cases." The CW 5G Debate in association with the NIC held at BT Tower, London, United Kingdom, November 8. https://www.cambridgewireless.co.uk/media/uploads/resources/Debates/08.11.16/5G_Debate2_08.11.16_FinalTranscript.pdf.

List of Abbreviations

3GPP	Third Generation Partnership Project
5GPPP	5G Private Public Partnership
AI	artificial intelligence
ARPU	Average Revenue Per User
BCG	Boston Consulting Group
CDMA	code division multiple access
cm	centimeters
DSSS	direct sequence spread spectrum
EBITDA	earnings before interest, tax, depreciation and amortization
eICIC	enhanced Inter-Cell Interference Cancellation
eMBMS	evolved Multimedia Broadcast Multicast Service
FCC	Federal Communications Commission
FDD	frequency division duplex
FTTC	fiber to the cabinet
FTTH	Fiber to the home
FTTP	fiber to the premise
FWA	fixed-wireless access
GB	gigabytes
Gbps	gigabits per second
GDP	gross domestic product
GPRS	general packet radio system
GSM	Global System for Mobile communications
HSPA	high speed packet access
IMT	International Mobile Telecommunications
IoT	Internet of things
LAA	licensed assisted access
LTE	long-term evolution (of cellular technology)
LTE-M	LTE for machines
M2M	machine to machine
MB	megabytes
Mbps	Megabits per second
MIMO	multiple-input multiple-output (antennas)
mmWave	millimeter wave bands
MNO	mobile network operator
MSC	mobile switching center
NB-IoT	narrowband IoT
NFV	network function virtualization

NGMN	Next Generation Mobile Network
OFDM	orthogonal frequency division multiplexing
OTT	over the top
RAN	radio access network
SE	spectrum efficiency
SMS	short message service
SNR	Signal-to-noise radio
SSID	service set identifier
TCP/IP	transfer control protocol/internet protocol
TDD	time division duplex
UHF	ultrahigh frequency
VAS	value-added services
VNF	virtualized network function
VNI	visual networking index
VoIP	Voice over internet protocol
VoLTE	voice over LTE
VR	virtual reality
V2V	vehicle to vehicle
WP-5D	Working Party 5D

Index

A

Access Wi-Fi, public 100
Additional capacity 78
Aircraft 78–79, 107
Airline Speed 13–14
Alcatel 12
Alternative Futures 91–92, 94, 96, 98, 100, 102, 104, 106, 108
Amazon 109–10
Analysts 26–27, 68
Analysys Mason 83, 86
Antennas 38–40, 51, 119, 123, 129
– additional 39–40
– beam-forming 51, 120
– directional 45, 120
– multiple 38, 40
– new 58, 61
Apple 27, 58, 110
Applications 16–19, 26, 28, 30–34, 46, 73, 77, 80–83, 91
– deployed 30
– envisioned 28
– machine 30
– new 27, 33–34, 80
Apps 16–18, 25, 70, 106, 108
AR (Augmented reality) 33–34, 74, 81
Architectures, configure network 48
Areas, suburban 120
ARPU Growth 59, 61
ARPUs 43, 53–55, 59, 62
Asian-Pacific countries 44, 46, 119
Asset tracking 30
AT&T 57, 103
Auction fees 53, 58
Auctions 54, 61, 110, 115
Augmented reality (AR) 33–34, 74, 81
Australia 18, 94, 116
Australian Government 94, 111
Authentication 31
Automated traffic control and driving 82
Automatic medicine dispensers 30
Automotive industry 30
Autonomous cars 27, 80, 83, 87–89

Autonomous operation 87
Autonomous vehicles 32, 83, 86–89, 91
Average mobile user 22, 32

B

Backhaul 44–45, 48, 93, 114
Backhaul connection 93
Bands 26, 40, 44–47, 51, 77, 115, 117, 119, 127
– guard 10, 116
– multiple 93
Bandwidth 33, 37, 93
Base station antennas 39, 41
Base station locations 42, 45, 120
Base station sites 40, 58
Base stations 38–40, 45, 47, 50–51, 58, 61, 93–94, 120–21, 123
– new 11, 94
Battery consumption 29, 45
Battery life 29, 79
BCG 17–19
Beams 38–40, 42, 45–46, 65
– optimal 40, 45
Bluetooth 28–29, 32, 83–84
Bodies, international 65
Broadband 77, 103, 108, 113–14, 116, 118, 120, 122
Broadband access 75–77
– dense area 76–77
Broadband access networks 35
Broadband fixed access 119
Broadcast 85
BT 67, 109
BT Wi-Fi 100, 103
Buildings 18, 28, 33, 39–42, 44, 92, 94, 96, 98
– public 92, 96, 104
Business Cases 51, 68, 85–86, 89, 128
Business models 114

C

Cambridge Wireless 46, 51, 79, 86, 125, 128
Capabilities 12, 18, 73, 78, 85, 115, 120

DOI 10.1515/9781547401161-012

Index

Capacity 10, 25–26, 38–39, 41–44, 46–48, 50–51, 62, 73, 92–93
- improving 85, 124
- increased 9, 25, 102
- road 88–89, 91
- sufficient 77, 79
Capacity and cost of small cells 43
Capacity cases, high 48
Capacity enhancements 40, 51, 123
Capacity gains 39, 41, 51, 68
Capacity growth 91
Capacity requirements 77
Capacity solutions, higher 73
Carriers 18, 127
Cars 31, 79, 83, 88–89, 91
CDMA (code division multiple access) 11, 129
Cell capacity 38, 40, 48, 98
Cell edges 39, 76
Cell phones 11, 49, 99–100
Cells 11, 38–41, 44–46, 48, 57, 78, 113, 121, 127
- smaller 38–39
Cellular communications 29, 113
Cellular coverage 93, 95–96, 98, 102
Cellular network equipment 126
Cellular networks 19, 30, 32–33, 47–49, 51, 85, 93
Change, regulatory 109
Channel 37, 39
China 16, 44, 46, 54, 117, 119
Circuit 11
Cisco 51, 126
Cities 44, 46, 61, 102, 118
Claims 65, 69, 80, 127
Clearwire 119
Cloud 50, 76, 78
Code division multiple access (CDMA) 11, 129
Com/globalmobiletrends 35, 71
Communications 31, 34, 45, 81, 87, 94, 98, 111, 124
- low latency and robust 78
Communications environment 113
Communications networks 50
Community 10, 19, 38, 56, 85, 87, 124
- cellular 108–9
Companies 12, 14, 25, 50–51, 53, 66–71, 100, 103, 109–10

- major 53, 69
- single 98, 101
Competition 53, 55, 58, 103–4, 109–10, 113–16, 118, 120–22, 125
- increased 59, 113
- reduced 114
Competitive disadvantage 71, 123
Competitive pressures 114, 116
Competitors 12, 59, 63, 67, 110, 123
Computing, remote 78
Computing platforms, general-purpose 38, 47
Congestion 48, 101
Connected machines 27–28, 31, 34
- world of 27, 30–32
Connections 38, 55, 59, 76, 82, 87, 91, 111, 113
Connectivity 16–17, 73, 78, 82–83, 87–89, 91, 101, 103, 110–11
Consistent connectivity 104, 108, 124, 127–28
Consumer-price connectivity package 55
Consumers 28, 30, 53–54, 59–60, 103, 110–11, 113–15, 124, 126–27
Consumers value 60
Contract 97, 103–4
Control 19, 28, 49–50, 80, 82–83, 88, 102, 127
Core network changes 50
Core network cost 48
Core network evolution 47
Core network upgrades 61
Core networks 19, 48, 50–51, 113, 125
- actual 48
Cost of small cells 43
Cost savings 48, 58
Costs 9–11, 47–50, 58–61, 77, 83–84, 86, 88–89, 93–94, 119–20
- consumer 103
- extra 51, 123
- high 43, 45, 51, 123
- low 29, 48, 111
- lower 77, 109–10
- lowest 54, 94
- operational 61, 121
Coverage 26, 29, 45–46, 77–79, 82–84, 86, 88, 92–95, 124–25

– better 49, 85, 94, 110, 113
– enhancing 18, 77
– global 30
– improved 84, 113
– poor 97, 125
– providing 41, 98
– rural 94, 97, 114
Coverage objectives 94, 105
Coverage obligations 94, 115
Coverage problems 92–93
CTIA 58, 63

D
Data capacity growth 25
Data growth 22–23, 25–26, 34, 108, 123
Data rate service, highest 18
Data rates 11–12, 15–17, 76, 85, 91, 94, 120, 124, 127
Data volumes 15, 20, 24, 27–28, 59, 74, 108, 119
Database 28, 107
Data-optimized networks 12
DC 63, 122
Degrees mobile 99, 111
Delay, radio 19
Demand, grow indefinitely 15–16, 18, 20, 22, 24, 26, 28, 30, 32
Dense areas 61, 75, 88, 92, 96, 102
– broadband access in 75, 77
Dense networks 52
Dense Wi-Fi deployments 77
Density, high 78–79
Deploy 44, 58, 60, 62–63, 67, 69, 93–94, 96–98, 119–20
Deploy indoors 44, 98
Deploy Wi-Fi 102, 105
Deployed unlicensed solution 106
Deploying Wi-Fi 107
Deployment 42–43, 68–69, 84–85, 93–94, 96–99, 106, 118–19, 121, 125–26
– dense 51, 123
– early 118–19
– practical 40
– slow 68, 117
Deployment costs 46
Device costs 79
Device manufacturers 29, 70, 108

Devices 22, 25–30, 32, 38–39, 48–49, 51, 78–79, 97–100, 102
– connected 29–30, 59, 74, 98
– consumer 30, 46
– mobile 17, 22
– multiple 49, 73
– short-range 29
– subscriber 38–39
Disaster situations 82
Downlink 47, 87
Download 15, 22, 83, 87, 97, 100, 102
Drones 27, 79, 82, 84
DSSS 94, 129
Duplex 47
– full 47

E
EBITDA 56
EC. See *European Commission*
Efficiency 15, 37–38, 74, 81, 101
– technical 37
EICIC 42
Electricity companies 28
EMBMS 84
Emergency service communications 81
Encryption 31, 101
Enhanced coverage 91, 97
– delivering 111, 124
Enterprise value (EV) 56, 63
Entrants, new 12, 51, 126
Equipment 31, 51, 61, 67, 69, 93, 117, 120
– new 66
Equipment manufacturers 110
Ericsson 12, 14, 20, 29, 35, 51, 66, 108, 122
Europe 10, 16, 19, 44, 46, 69, 71, 117–18, 126–27
European commission (EC) 61, 65–66, 69, 84–86, 96, 107, 114, 118
European Union 63, 115
EV (enterprise value) 56, 63
Executive officers 68, 106, 110
Expenditure 57, 59
Expense 79, 94, 115

F
Facebook 34, 50, 58, 109
Failure of Municipal Wi-Fi 102

Fastest Networks 86
FCC (Federal Communications Commission) 54–55, 57, 63, 110, 113, 115
FDD (frequency division duplex) 47, 129
Features, new 27, 50
Federal Communications Commission. See FCC
Femtocells 44, 70, 95, 98
Femtocells and small cells 96
Fiber 18, 37, 48, 119
FinalTranscript.pdf 51, 86, 128
Fixed-wireless access. See FWA
Forecast period 59
Forecasts 21, 23, 25, 27, 32, 34, 106
Frequencies 10, 41–42, 44, 73, 115–17
– high 116–17
Frequency bands 11, 39, 46–47, 65, 102, 116–17, 119, 121
– new 51, 123
Frequency division duplex (FDD) 47, 129
FTTC 18, 104
FWA (fixed-wireless access) 45, 107, 109, 119–22, 125, 129

G
GGeneral Packet Radio Service (GPRS) 10, 129
Generation Mobile Networks 86
Generations 11–12, 25, 37, 50, 58, 66, 113–15, 117
– first 9
– previous 9–10, 12, 51, 65, 73, 85, 116, 123–24
Germany 33, 65, 86, 116
G.fast 18, 104, 120
GHz 26, 40, 44, 46, 61, 77, 115–16, 119, 121
Gigabytes 22, 32, 53
Global mobile trends 35, 63, 68, 71
Google 49–50, 58, 70, 83, 87–88, 103, 106–7, 110, 125
Government buildings 102–3, 107
Governmental action 103
Governments 68, 94, 96–97, 100, 102–4, 106–8, 113, 124, 127
Governments and regulators 103, 111, 117, 124
GPRS (General Packet Radio Service) 10, 129
Grid networks, smart 80

Growth 10, 15, 23, 25–26, 32, 34, 37, 54–55, 62
– fast 25, 35
– rapid 11, 29
– subscriber 53–54, 59
GSM 10, 119
GSM Global System for Mobile communications 129
GSM system 81
GSMA 26, 33, 35, 54, 56–57, 63, 68–69, 71, 110
GSMA Intelligence 35, 63, 71

H
Hackers 9, 100
Hacking in 31
Handset subsidy 58, 61
Handsets 17, 19, 47, 49, 92, 99, 102, 106, 127
Harmonization 118
– global 46
HD (high definition) 16, 74
Heaviest users 25
Heterogeneous Networks 48–49
Hetnets 49–50
Hich-networks-offer-wi-fi-calling 111
High data rates 10–11, 76, 104
High definition (HD) 16, 74
High speed packet access. See HSPA
High speed trains 78, 92
Higher data rates 10, 12, 15, 18, 82, 94, 102
Home automation 28
Home broadband 104
Home devices 31
Home network, correct 31
Hong Kong 23
Hotspots 41, 43
Hour MB 43
HSPA (high speed packet access) 11, 129
HTC 12
HHuawei 12, 14, 66–67, 108

I
Identification 63, 100
IEEE Communications Magazine 51–52
IMC (Internet Measurement Conference) 35
Impact on Spectrum and Network Use 86
Implications for Network Load 32

Index — 135

IMT (International Mobile Telecommunications) 74, 86
India 16, 54
Indoors 43–44, 46, 76, 81, 102
– deploying cellular 98
– trialed 33
Industry Structural Change 125
In-home network solutions 29
Innovation 104, 113, 115, 118, 121, 125–26
Innovation Centre 65–66, 71
Integration 49, 102, 104
Interference 11, 40–44, 47, 80, 93, 95–98, 101
International Mobile Telecommunications (IMT) 74, 86
International Telecommunication Union (ITU) 74
Internet 10, 19, 27–29, 31, 78–79, 105
– tactile 33, 80–81, 85
Internet connectivity 96, 111
Internet Measurement Conference (IMC) 35
Internet of things. See IoT
Internet protocols 16, 130
Investment 9, 18, 56, 58, 60–63, 66, 104, 115–17, 123–26
Investment Choices 60
Ionica 119
IoT (internet of things) 27–29, 31–32, 78–80, 85, 105–8, 127, 129
IoT connectivity 79
IoT devices 49, 84
IoT networks 48
IoT solutions 79, 105
IoT technologies, emerging 86, 124
iPhone 11, 20, 27, 29–30, 70
ITU (International Telecommunication Union) 74

J
Japan 10, 23, 44, 119, 126

K
Key players 65–66, 68, 70, 109, 123, 126, 128
Korea 46, 119
Korea Telecom 67, 126

L
LAA (licensed assisted access) 102–3, 129
Latency 11, 16–20, 74, 80–82
– low 73, 76, 78, 80–82
– lower 20, 74
– reduced 74, 127
Leadership 67, 69, 118, 122
Licensed assisted access. See LAA
Limits of small cells in dense networks 52
Listed companies 53, 123
LMDS (local multipoint distribution service) 119
Load 19, 32
Local multipoint distribution service (LMDS) 119
Locations Wi-Fi 111, 124
Logistical problems 93–94
London 51, 61, 63, 86, 128
LoRa 32, 79, 105
Lower cost provision of basic services 76
Low-power networks, single 49
LTE 37, 79, 119, 129–30
Lucent 12

M
Machine connectivity 29–31
Macrocell 39–44
Maintenance depots 93
Major Players 70
Malls 42, 97
– shopping 41–42
Manifesto for timely deployment 69, 71, 122
Manufacturers 25, 46, 66–67, 70, 74, 115, 119, 123, 126–27
– major 65–66
Manufacturers of cellular network equipment 126
Markets 16, 29–31, 46, 53, 59, 69, 84, 106, 108–10
Masts 50, 94, 114
Mergers 58, 107–8, 113–14, 121–22
Message service, short 10, 130
Metallized windows 92
Meters 31, 40–41, 45–46, 113, 120
– smart 28–29
Microcells 39, 41–43
Millimeter wave bands 40, 73

MIMO (multiple input, multiple output) 38–40, 129
MIMO, classic 38–40
MIMO deployment 39
MmWave 40, 44–46, 51, 65, 69
MmWave bands 44, 116
MmWave deployments 46, 61, 77, 81
MmWave frequencies 68, 115
MNO mobile network operator 129
MNO Performance 53, 55, 57
MNOs 25–27, 48–51, 53–54, 58–63, 67–70, 91–96, 102–10, 113–17, 125–27
– limiting 9
– major 110
– multiple 44
– native 53
– rational 62
– resident 53
MNOs building 81
MNOs deploy 126
MNOs hedge 60
MNOs in rural areas 97
MNO's network 95, 99
Mobile 35, 38–40, 47, 50, 73, 86, 103, 115, 127
Mobile communications 9–10, 44–45, 103, 105, 113, 128–29
Mobile community 85, 124
Mobile coverage 17, 94
Mobile data 20, 29–30
Mobile handset users 16
Mobile industry 62, 116, 123, 125
Mobile network operators 43, 67
Mobile networks 10, 35, 38, 47–48
Mobile operators 39, 94, 120
Mobile payment 70
Mobile phone users 107
Mobile phones 9–10, 17–18, 25, 47, 108, 124, 128
Mobile radio systems 41, 81
Mobile sector 115, 121, 124
Mobile services 15, 77
Mobile switching centers (MSCs) 47
Mobile technology 15, 78, 113
Mobile users 12, 44, 91
Mobile World Congress 28, 110
Modes 81, 95, 100–101

Mogensen 37, 51
Monitoring 30–31, 101
Motorola 12, 119
Move Wi-Fi 97
Ms delay 19
MSC mobile 129
MSCs (mobile switching centers) 47
Multiple input, multiple output. See *MIMO*
Municipal Wi-Fi 102

N

NB-IoT 79, 105, 108, 127
NB-IoT narrowband IoT 129
Netflix 16, 35
Network architecture 28
Network capabilities 50
Network capacity 12, 26
– higher 44
– total 48
Network congestion 18, 84
Network connectivity 83, 88
Network connectivity help 88
Network control 87–88
Network cores 38
Network cost 47
– total 48
Network coverage 78, 83
Network design 65
Network enhancements 33, 123
Network equipment, new 54
Network expansion 110
Network function virtualization. See *NFV*
Network functions 48, 81
– virtualized 130
Network generation 53
Network investment 59
Network level 49
Network Load 32
Network loading 34
Network operation 81
Network operators 66
– virtual 50
Network outages, complete 48
Network problems 48
Network protocols 50
Network slicing 48, 50
Network software 29

Index — 137

Network traffic 33
- total 81
Network usage 48
Network use 86
Networks 9–11, 17–20, 33–34, 45–49, 62–63, 84–89, 100, 114–16, 127
- centralized 88
- current 26
- external 93
- interconnected 28
- loaded 80
- low-cost 76
- low-latency 89
- multiple 29, 79
- nationwide 114
- reliable 76
- sensor 79
- single 46, 48, 114
- social 76
- software-defined 102
- suboptimal 33
- video-dominated 34
New services 16, 22, 46, 48, 54, 62, 73, 85
Next Generation Mobile Networks. See *NGMN*
NFV (network function virtualization) 48, 50–51, 102, 130
NFV network function virtualization 130
NGMN (Next Generation Mobile Networks) 73, 75, 82, 85–86
NGMN Alliance 73, 75, 79, 82
NGMN Next Generation Mobile Network OFDM 130
NGMN View of Potential Applications 75, 77, 79, 81, 83
NIC 51, 86, 128
Nokia 12, 51, 66, 108, 119

O

Ofcom 113, 122
OFDM (orthogonal frequency division multiplexing) 11, 98
Office 28, 32, 76–77, 92, 96, 98, 103–4, 107
Open Wi-Fi framework, public 107
Operator cloud services 76
Operators 46–47, 53, 59–61, 65, 84, 86, 99–100, 105–6, 109
- single 46, 106

Operators deploy 127
Operators fear 71, 123
Orthogonal frequency division multiplexing (OFDM) 11, 98
OTT providers 50, 125–26
OTT service 50

P

Page Load Times 17
Password protection 99–100
Passwords 31, 95–97, 99–101, 103
Path 38–39, 71, 106
Peak data rates 74
Performance Bottlenecks in Broadband Access Networks 35
Phones 9, 11, 49, 70, 84, 103, 124
Picocells 44
Picture messaging 70
Plateau 22–23, 25–26, 34, 77, 123
Policies 101, 103–4, 113, 124–25
Portugal 23
Precision network connectivity of cars 89
Predictions 13, 20–21, 25, 32, 37, 54, 59, 71, 105–6
Predictions of volume 20–21, 23, 25
Prices 23, 59, 62, 108–9, 113–14, 119–20
- lowest 33, 115
Prior Generations 10, 12, 14, 73
Prisoner's dilemma game theory 60
Privacy 31
Productivity 9, 31, 68, 91
Profitability 53, 67, 108, 115, 125–26
Proposal 111, 114
Proprietary IoT networks 32
Providers 50, 103–4, 125
- connectivity platform 109–10
Provision 76, 78, 83, 100, 105
- universal service 103

Q

Qualcomm 11–12, 118

R

Radio access 38, 47
Radio access networks. See *RANs*
Radio energy 39–40
Radio paths, multiple 38–39

Radio solution, new 38, 61
Radio spectrum 15, 116
Radio technology 86, 124
Railway network operators 93
RANs (radio access networks) 50, 125–26, 130
Rates 17–18, 33, 53, 59
R&D 12, 53, 69
Realwire 23–24, 35
Receiver 38, 40
Refarming 117
Regulation 101, 103, 105, 109, 111, 113–16, 118, 120, 124–25
Regulators 25, 65, 67–68, 100–101, 103, 107–11, 113–18, 124–25, 127
Reliability 82, 113
– high 82–83
Repeaters 92–93, 95
Research 12, 40, 46, 51, 65–66, 69, 110
Resolution video services, higher 15
Resource blocks 19, 42
Revenues 50, 53–55, 59–60, 62–63, 70, 110, 115, 123, 125
– annual 70
– average 53, 61
– increased 85
Risk 33, 38, 48, 67, 109
Roaming 9–10
Robots 27, 83
Robust communications 78
Role 12, 50, 66, 79, 84, 101, 121–22, 124–25
Rollout 30, 56, 61, 69–70, 123
Rosenthal 16–17, 35
Routers, government Wi-Fi 108
Rural areas 75, 92, 94, 97–98, 111, 120, 124
Rural network, single 104

S
Safety 88
– public 81–82
Sales, mobile network infrastructure 108
Samsung 12, 66–67
Satellite 87, 93, 103, 115
SDN (software-defined networks) 102
Sectors 43, 56–57, 63, 67, 114–15, 125
Security 9–10, 28, 31, 82, 101, 105
Self-deployed Wi-Fi 95

Sensors 28–29, 32, 78, 88–89
Server 16, 19, 101
Services 48–50, 54, 60–61, 66–68, 70, 73, 84, 107–10, 124–27
– better 49, 59
– delivering 10
– emergency 48–49, 81–82
– low-latency 81
– value-added 59, 76, 130
– voice-over-Wi-Fi 49
Shannon's limit 37
Shared network access 104
Shared networks 50, 104, 114, 116, 121
– single 125
Siemens 66
Sigfox 32, 79, 105
Signal strength 41
Signals 39, 70, 73, 78, 87–88
Singapore 23–24, 107
Skype 70, 93, 99
Small cell deployments 44
Small cells 38, 41–46, 51–52, 58, 61, 68, 96, 102, 123
– additional 43, 123
Social networking 34
Software-defined networks (SDN) 102
Solutions 30–31, 38, 76–77, 79–80, 83–84, 94–97, 99, 119–20, 126–27
– better 111, 124
– fastest 67, 127
– low-latency 45, 109
– mmWave 45–46, 115
– new 38, 120
– unlicensed 79, 106
– utopian 85, 124
– voice-over-Wi-Fi 93
Spectrum 26, 37–38, 42, 44, 53, 86, 98, 113–15, 117–18
– licensed 101–2, 105
– new 10, 26, 58, 116
Spectrum acquisition costs 61
Spectrum efficiency 11, 37, 44, 48, 81, 130
Spectrum position 117
Spectrum regulators 113
Speeds 9, 14–15, 17–18, 91, 93, 119, 126
– current mobile data 34, 123
– high 16, 18, 67, 78, 91

– higher 9, 15, 34, 102, 123
– lower 16–17
Spots, hot 42, 78, 100, 103, 108
SSIDs 97, 101
Standards 12, 19, 65, 69, 94, 118
Starbucks 97, 103
Subscriber base 27, 54, 58, 61
Subscriber churn 58, 60, 63, 123
Subscriber costs, direct 61
Subscribers 15, 18, 53–54, 59–63, 67, 109, 119, 123–24, 126
Subsidizing 94
Support 10–11, 46, 66, 68, 70, 73, 99, 114, 118
Surrey 66, 71
Sweden 23–24, 35
Switch 60, 95, 120

T

TCP/IP transfer control protocol/internet protocol TDD time division 130
TDD (time division duplex) 26, 130
Tech4i2, 61, 63
Technology 9–12, 34, 37–38, 50–52, 77–78, 80, 85–86, 117–19, 123–27
– better 10, 12
– cellular 10, 116, 129
– digital 9–10
– new 27, 63, 67, 69, 79, 118–19, 121, 123, 126
– wireless 30, 80
Technology readiness 51, 86, 128
Tefficient 23, 35
Telecoms 63, 71
Telefonica 114
Telemetry data 83
Test-beds 65, 104, 118
Time division duplex (TDD) 26, 130
Towers 94
Traffic 17, 27, 32, 42–44, 47–49, 76, 81, 92, 98
Train stations 97
– major 92, 96–97, 109
Trains 92–93, 104–5, 107, 111, 124–25
Transmissions 32, 47–48
Tunnels 78, 92–93, 99
TV 120

TVs, connected 30

U

UK 63, 86, 122
UK National Infrastructure Commission 61, 63
UK networks Ooffer Wi-Fi calling 111
UK's BT Wi-Fi 97
Ultralow latencies 33, 65
Ultralow latency solutions 126
Uncongested networks 18
United Kingdom 51, 61, 65, 86, 98–99, 105, 109, 114, 128
United States 10–11, 19, 44, 46, 56–58, 115–16, 118, 120–22, 126–27
Unlicensed spectrum 26, 101–2, 105
Upgrade 18, 50, 58, 63, 66, 123, 127
– business as usual network 57
Uplink 47, 84
Urban areas, dense 61, 89, 92, 102
US Average Revenues 55
US companies 12, 122, 125
Usage cases 85, 119
Use Cases 77–79, 82
Users 9–11, 15–17, 22–23, 38, 40–41, 44–45, 49–50, 93, 97–101
– authenticate 100
– improved 73
– indoor 43, 45
– multiple 37
– potential 80
– reason 103

V

VDSL 18, 104
Vendors, traditional 51
Verizon 57, 103, 119, 121
Video 10, 15–19, 22–23, 25, 33, 70, 78, 82, 121
– download 22, 83
– hour of 22, 32
Video streaming 15, 18, 22, 121
Video telephony 11
Virtual networking index 20
Virtual reality. See *VR*
Virtualization, network function 48
Visual networking index 130

VNI 20, 130
Vodafone 53, 67
Vodafone CTO 71
Voice 10, 12, 20, 44, 53, 93, 99, 104, 108
Volume 15, 20–21, 23, 25, 89
– high 76
VR (virtual reality) 16, 19, 33, 80–81, 130

W, X, Y
Washington 63, 122
Wavelengths 39–40, 44
Web browsing 18–19
Web pages 19
Webb 42–43, 52, 86, 120, 122
Website 71, 111
Weightless 32, 79, 105–6
WhatsApp 59, 93, 99, 108–9
Wide area devices, high-value 29
Wide area network 28
Widgets 70
Wi-Fi 17, 25–26, 28–29, 49–50, 59, 77–78, 95–105, 107–11, 124–26
Wi-Fi access 99
Wi-Fi access enabler 103–4
Wi-Fi access points 17, 50, 95, 98
Wi-Fi access points by myriad entities in unlicensed spectrum 105
Wi-Fi availability, increased 108
Wi-Fi changes and additions 99
Wi-Fi connection 99–100
Wi-Fi connectivity 59, 98
Wi-Fi coverage 95, 102

Wi-Fi deployment 107
Wi-Fi device 101
Wi-Fi first policy 101
Wi-Fi First World 97, 99, 101, 106, 111, 124
Wi-Fi frequencies 109
Wi-Fi networks 102–4
– near-ubiquitous public 107
– universal 104
Wi-Fi nodes 101
Wi-Fi password 99, 109
Wi-Fi playing 125
Wi-Fi projects 107
Wi-Fi providers 105
Wi-Fi provision 25, 48
Wi-Fi repeaters 87, 92–93
Wi-Fi router 97
Wi-Fi router owner's network 100
Wi-Fi signals 92
Wi-Fi solutions 44, 46, 96, 104
Wi-Fi spectrum 17, 109
Wi-Fi usage 106
Wi-Fi2, 96
Wi-Fi-connected phones 105
Wi-Fi-oriented MVNOs 107
Wi-Fi-related initiatives 106
Wired home broadband services 119
Wireless 56, 63, 87, 113, 119, 122, 128
WP-5D 74

Z
Zigbee 28

Our Digital Future

A new book by William Webb—available from Amazon

Alluring visions of the future abound, yet flying cars have not filled our skies, and the smart fridge is more of a joke than a reality. But digital technology has changed our lives completely—for better and for worse—with always available connectivity, Facebook, Uber, and so much more. With jobs potentially at threat and political instability rising, correctly predicting our digital future is more important than ever. Dr. Webb has had an outstanding track record of forecasting over the last 20 years and applies the same pragmatic realism to the next 20.

Chapter 1 begins by looking at what is and what is not included. This book is looking at the impact of digital technology on the future, so areas such as mobile communications are included, but areas such as advances in medical care are not.

Chapters 2 and 3 discuss other social influences. Chapter 2 looks at books on magic and on science fiction to see what their authors imagined in a world where anything might be possible. This provides a guide to what we desire. Chapter 3 looks at previous predictions to see if there are systematic errors or other biases that might be avoided. It concludes that until about 50 years ago we tended to underpredict the future, but now we have swung the other way and are often much too optimistic.

Chapter 4 looks at areas of technology where there might be rapid improvements and, in particular, those which might enable progress across many areas—as, for example, the iPhone has. It concludes that key future enablers will include the internet of things (IoT), artificial intelligence, and perhaps robotics.

The next five chapters look at different environments, such as the home, work, leisure, and public services. These chapters show that in some cases (such as the home and public services), change will be slow and minimal, whereas in others (such as work and leisure), much more significant changes can be expected.

Chapter 10 asks whether societal issues might have an impact. There is a growing backlash against some aspects of the digital world, and this might slow the pace of change. The chapter concludes that although there is disquiet, it will be insufficient to make much of a difference, except in specific cases such as terms of employment.

Chapter 11 then brings it all together with a set of predictions looking 10, 20, and 30 years from now. Most of the future enablers are associated with business rather than the individual. Hence, the changes noticed by the individual may be relatively small compared to the changes of the last 30 years. Individuals will see even better virtual assistant functionality from their devices, as solutions such as Siri steadily improve using emerging artificial intelligence (AI) techniques. In the home, some new connected devices such as smart speakers and home IoT products will be installed, but home automation will not improve much. Leisure interests will expand, with each genre (e.g., cycling) gaining apps, online communities, additional functionality, and, where appropriate, monitoring from IoT devices. This will allow us to spend more time on our favorite pastimes, as indeed we may need to if enhanced productivity and automation leads to fewer jobs.

In business, the office will see widespread deployment of IoT, biometrics, and robotics, mostly as a way to save costs on administrative and maintenance staff. Some sectors will make extensive use of the IoT to improve productivity in such areas as agriculture and manufacturing. Some markets (such as retail) will decline further due to changing habits. Some industries, such as construction and hospitality, will be broadly unaffected. Vehicle maintenance, which is currently a huge employer, may decline as more electric vehicles are introduced.

Transportation will not change much. Improvements to be seen regarding transport will be better connectivity while traveling, having better access to trip information, and having a gradual growth in driverless vehicles (cars, trains, buses, etc.).

In essence, the key gains will be in convenience, productivity, and reliability. The world will be a similar place to today but will work better.

Finally, Chapter 12 looks at the impact that our possible future world might have on the structure of industry. It predicts that today's large digital companies, such as Google and Amazon, will continue to dominate well into the future. New players such as Uber and Tesla will emerge but at a slower rate. A few, such as Facebook, might struggle as regulation strengthens. Connectivity providers (e.g., mobile network operators) will become utility-like, and therefore their manufacturers will struggle.

Printed in Great Britain
by Amazon